TURBO CODING FOR SATELLITE AND WIRELESS COMMUNICATIONS

THE KLUWER INTERNATIONAL SERIES IN ENGINEERING AND COMPUTER SCIENCE

TURBO CODING FOR SATELLITE AND WIRELESS COMMUNICATIONS

M.R.SOLEYMANI
Concordia University

YINGZI GAO
Concordia University

U. VILAIPORNSAWAI
McGill University

Springer Science+Business Media, LLC

Additional material to this book can be downloaded from http://extras.springer.com

ISBN 978-1-4757-7658-4 ISBN 978-0-306-47677-8 (eBook)
DOI 10.1007/978-0-306-47677-8

Library of Congress Cataloging-in-Publication Data

Turbo Coding for Satellite and Wireless Communications
M. R. Soleymani, Yingzi Gao and U. Vilaipornsawai

Originally published by Kluwer Academic Publishers in 2002
Softcover reprint of the hardcover 1st edition 2002

Printed on acid-free paper.

Contents

List of Acronyms

2D	Two dimensional
3D	Three dimensional
3GPP	3rd Generation Partnership Project
8PSK	8-ary Phase Shift Keying
A/D	Analog to Digital converter
ALD	Augmented List Decoding
APP	A Posteriori Probability
ARQ	Automatic Repeat reQuest
AWGN	Additive White Gaussian Noise
ASIC	Application Specific Integrated Circuit
ASK	Amplitude Shift Keying
BCH	Bose-Chaudhuri-Hocquenghem code
BER	Bit Error Rate
BCJR	Bahl-Cocke-Jelinek-Raviv
bps	bit per second
BPSK	Binary Phase Shift Keying
BSC	Binary Symmetric Channel
BTC	Block Turbo Code
BWA	Broadband Wireless Access
CCSDS	Consultative Commitee for Space Date System
CDMA	Code Division Multiple Access
CITR	Canadian Institute for Telecommunications Rearch
CPLD	Complex Programmable Logic Device
CPM	Continuous-Phase Modulation
CRSC	Circular Recursive Systematic Convolutional
CSA	Canadian Space Agency
D/A	Digital to Analog converter
DAB	Digital Audio Broadcasting
DAMA	Demand-Assigned Multiple Access
DSP	Digital Signal Processing
DVB-RCS	Digital Video Broadcasting-Return Channel via Satellite
DVB-T	Digital Video Broadcasting-Television

EEP	Equal Error Protection
ETSI	European Telecommunications Standards Institute
FEC	Forward Error Correction
FER	Frame Error Rate
FPGA	Field Programmable Gate Array
GF	Galois Field
GTPC	Generalized Turbo Product Code
HCCC	Hybrid Concatenated Convoultional Code
IP	Intellectual Property
LAN	Local Area Network
LDPC	Low Density Parity Check code
LLR	Log-Likilihood Ratio
MAN	Metropolitan Area Network
MAP	Maximum a posteriori Probability
MF-TDMA	Multi-Frequency Time-Division Multiple Access
ML	Maximum Likelihood
MPEG	Moving Picture Experts Group
M-PSK	M-ary Phase Shift Keying
MSB	Most Significant Bit
PAM	Pluse Amplitude Modulation
PCCC	Parallel Concatenated Convolutional Code
PCTCM	Parallel Concatenated Trellis Coded Modulation
PSK	Phase Shift Keying
QAM	Quadrature Amplitude Modulation
QPSK	Quadrature Phase Shift Keying
RCST	Return Channel Satellite Terminal
RM	Reed-Muller code
SNR	Signal to Noise Ratio
RS	Reed-Solomon code
RSC	Recursive Systematic Convolutional
SCCC	Serial Concatenated Convolutional Code
SCTCM	Serial Concatenated Trellis Coded Modulation
SISO	Soft-Input Soft-Output
SOVA	Soft-Output Viterbi Algorithm
SSPA	Solid State Power Amplifier
TCC	Turbo Convolutional Code
TCM	Trellis Coded Modulation
TCT	Time-solt Composition Table
TPC	Turbo Product Code
TTCM	Turbo Trellis Coded Modulation
UEP	Unequal Error Protection
UMTS	Universal Mobile Telecommunication Service

VA	Viterbi Algorithm
VSAT	Very Small Aperture Terminal

List of Figures

List of Tables

Preface

The introduction of Turbo codes in 1993 was evidence of the attainability of the error correction performance bounds derived by Shannon in 1948. The original turbo codes consisted of two recursive convolutional codes concatenated in parallel and decoded using an iterative message passing algorithm consisting of two *Maximum a posteriori Probability* (MAP) decoders. The astounding performance of these codes resulted in a surge in the research activity in the area of concatenated codes and iterative decoding techniques. The idea was soon extended to other codes and code combinations as well as iterative schemes using different techniques in their iterations.

The general nature of the *message passing* technique used for the decoding of turbo codes, i.e., the iterative exchange of soft information between two processing blocks, is now widely recognized as a very general and powerful concept whose applications go far beyond the decoding of these codes.

The material presented in this book is the result of the research conducted at the *Wireless and Satellite Communications Lab.,* Concordia University. In order to make the book self-contained, we have added the necessary background material. As our audience, we had in mind graduate students conducting research in the area of digital communications as well as the practicing engineers involved in the design of communication circuits and systems.

Our objective is to give the reader enough information enabling him/her to select, evaluate and implement the code suitable for his/her application. The programs in the CD-ROM and related material in the book can be easily used by the reader for simulation and performance evaluation of turbo codes.

The organization of the book is as follows. Chapter 1 serves as an introduction to the rest of the chapters. Chapters 2, 3 and 4 relate to turbo codes using convolutional codes as their building blocks. Chapters 5, 6 and 7 discuss *Block Turbo Codes* (BTCs) , i.e., turbo codes having block codes as their constituent codes. Chapters 8 deals with the issues concerning the implementation of turbo codes.

Another important class of linear block codes, *Low Density Parity Check* (LDPC) codes, invented in the early 1960s, has received considerable attention after the invention of turbo codes. With iterative *message-passing* decoding algorithms, variants of the LDPC coding techniques have exhibited a performance comparable to, and sometimes even better than, the original turbo codes. Chapter 9 of the book is devoted to this topic.

The work presented in this book would not have been possible without a research grant from the *Canadian Space Agency* (CSA) and the *Canadian Institute of Telecommunications Research* (CITR) entitled *Spectrum Efficient Transmission with Turbo Codes for Satellite Communication Systems.* The authors wish to thank the CSA and the CITR. They are particularly indebted to Dr. Birendra Prasada, the former president of the CITR, for his continued support, encouragement and constructive criticism. We would also like to thank NSI Global Inc. for their financial and technical support of the CSA/CITR project. The CSA/CITR project gave us the opportunity to collaborate with other researchers working on Turbo codes. We would like to express our appreciation for the fruitful interaction with John Lodge, Paul Guinand and Ken Gracie from the *Communications Research Centre* (CRC), A.K. Khandani of Waterloo University and F. Labeau of McGill University.

We would like to thank both the faculty and student members of the *Wireless and Satellite Communications Lab* for many helpful comments and suggestions. More than anyone else, we are grateful to Prof. J.F. Hayes for his active involvement in technical discussions with the authors, his suggestions for improving our simulation methods and proof-reading parts of the manuscript. We are thankful to Prof. A. Al-Khalili for his helpful comments on implementation issues. We would also like to thank Dr. Li Xiangming for many helpful comments.

We are most grateful to Mohsen Ghotbi for always being ready to lend us a helping hand and for proof-reading the final version of the manuscript, Bo Yin (now with PSQ Technologies Inc.) for her contribution to the programs for the simulation of Turbo Block Codes, Pourya Sadeghi for providing us with the program for the simulation of 3GPP (*3rd Generation Partnerships Project*) turbo codec.

The first author wishes also to express his gratitude to Dr. N. Esmail, Dean of the Faculty of Engineering and Computer Science for his continued support of his research and for awarding him the Concordia Research Chair in Wireless Multimedia Communication enabling him to intensify his research activity. He would also like to acknowledge the support received from the Natural Science and Engineering Council (NSERC) in the form of the Operating Grant OGPIN 001 for the past 14 years.

M. R. SOLEYMANI
Y. GAO
U. VILAIPORNSAWAI

MONTREAL, QUEBEC

This book is dedicated to our families

Chapter 1

INTRODUCTION

The publication of Shannon's historical paper[1] ushered in the era of reliable information transmission. The fact that Shannon's bounds could only be approached *asymptotically,* however, was conceived, until recently, as an indication of the unattainability of these bounds. Also, the proof of the channel coding theorem being based on a *random coding* argument led the coding theorists to believe that a good code (in the sense of achieving the channel capacity), should lack any structure [2] and, therefore, be almost impossible to decode. In the early 1990s, major advances in the area of digital hardware design, had made the implementation of some of the most complex functions feasible. These advances in digital electronics prompted some coding theorists to revisit the concepts of complexity and randomness [3] and others to look for practical decoding schemes for capacity achieving codes [4], [5]. However, it was not until the invention of Turbo Codes [6] and the demonstration of their amazing performance that the coding community's perception of *randomness, asymptotic* and *complexity* changed [1] and an intense research activity on iterative decoding of concatenated codes was initiated [7], [8], [9], [10], [11], [12], [13], [14].

An interesting aspect of Turbo codes is that their decoder was designed prior to their encoder [16], [17]. Earlier codes such as BCH and *Reed-Solomon* codes, were first developed based on mathematical (algebraic) principles, generally, without much attention being paid to their decoding complexity. The decoding procedure for these codes were discovered later. The same is true of the convolutional codes [18]. In the case of Turbo codes, however, the decoding structure, viz., the *turbo* or *message passing* decoding was designed first and the encoder implementation followed [17]. The original Turbo code [6] involved the parallel concatenation of two convolutional codes whose astounding performance was a major factor in the popularity of turbo codes. Soon after, it became clear that many other codes and code combinations can be used with

[1]For a very insightful summary of the evolution of the coding community's assessment of the decoding complexity of capacity achieving error correcting codes see R.J. McEliece's 2001 ISIT Plenary Lecture [15].

the *message passing* decoding. The message passing concept has also been used in many other communication problems where a detection problem can be broken into two or more coupled problems [19].

In this book, we have tried to present different turbo coding schemes, particularly, those suitable for wireless and satellite communications, in a unified framework. This unified framework is the result of our looking at these different code configurations from the decoding point of view and trying to adapt the same decoding procedure to all of these configurations. While we discuss different message passing schemes, e.g., different forms of *Maximum a posteriori Probability* (MAP), *Soft Output Viterbi Algorithm* (SOVA) and list decoding algorithm (Chase Algorithm) [10], our emphasis is on a particular flavor of MAP given by Hagenauer et al. [7]. Another point worth mentioning is that while some authors make a distinction between the original turbo codes, or at least, the parallel concatenated convolutional codes and other encoder configurations, we use the term *turbo code* to refer to any encoder configuration used in conjunction with an iterative message passing decoder.

As our audience, we had in mind graduate students doing research in the area of digital communications as well as practicing engineers involved in the design of communication systems. We assume that the reader has taken a senior level undergraduate course in digital communications and is familiar with the topics presented, e.g., in [20], [21].

Our objective has been to give a comprehensive treatment of the topic in order to enable the reader to select, evaluate and implement the turbo code suitable for his/her application. the programs in the accompanying CD-ROM can be easily used by the reader for simulation and performance evaluation and serve as a good starting point for a design effort.

We were tempted to delete some of the extra steps in the derivation of the decoding algorithms in order to improve readability. However, we finally decided to keep these details believing that they may be of use to those readers who want to simulate or implement the algorithms. Those readers not interested in these details or those finding them trivial may skip them.

In this chapter, we briefly present some basic concepts from information theory and coding theory that we consider a useful introduction to the rest of the book. Readers familiar with these concepts, may skip the rest of this chapter.

1.1. Error Control Coding

Figure 1.1 is a block diagram of a communications link. Here, we assume that the output of the source is time-discrete and takes a finite number of values. This means that the information is either digital to start with, e.g., the output of a digital processor, or has been subject to sampling and quantization. The output of the source is first encoded using an error control coding scheme. This involves the addition of some redundant symbols to a group of source symbols. The encoded data stream is then modulated and sent over the channel. The object of modulation is to turn the encoded symbols into signals better suited for transmission over the channel. The effect of the channel is corruption of the signals through the addition of noise and other artifacts. At the receiver side, the received noisy signal is first demodulated recovering the encoded symbols, possibly with error. Then, the decoder attempts to correct the errors using the extra information available thanks to the redundancy added by the channel encoder.

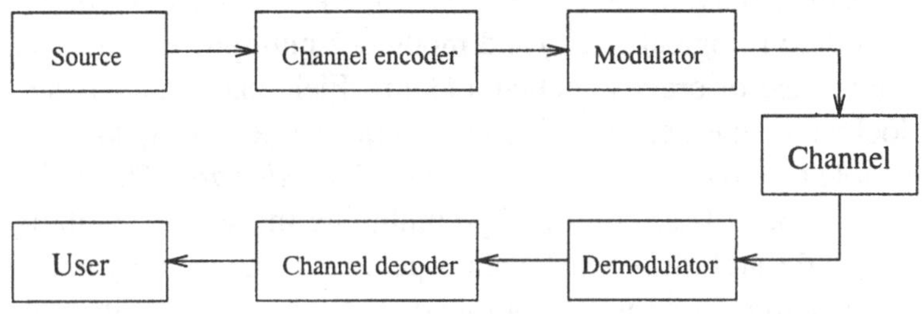

Figure 1.1. Block Diagram of a Communications Link

The above arrangement is usually referred to as *hard decision decoding*. A more efficient decoding approach is to combine the demodulation and decoding functions, i.e., to pass the output of the channel directly to the decoder. In this scheme, called *soft decision decoding,* the decoder has access to more information about the transmitted data and, therefore, better performance is achieved [22]. To further improve the transmission quality the functions of the modulator and the channel encoder can be combined. In this approach, called *coded modulation* [101], instead of devising the encoding and modulation schemes separately, the code design is related to the given signal constellation.

Error control coding schemes, in general, can be divided into two broad categories of *Automatic Repeat Request* (ARQ) and *Forward Error Correction (FEC)* codes. An ARQ system detects the errors and asks for retransmission of erroneous packets while in systems using FEC, the decoder tries to correct as many errors as possible. Since error detection requires less redundancy, ARQ is more bandwidth efficient (requires less overhead) when the channel is good and its efficiency deteriorates gradually with the channel condition. For a delay sensitive application, the performance of the ARQ (in terms of delay between consecutive packets) deteriorates as a function of the distance and the transmission speed (bandwidth). The reason is that an increase in any of these two parameters results in an increase in the delay (for the receiver to request for retransmission) relative to the duration of a packet. As a result, the use of ARQ is very limited in the communication systems where a large portion of the traffic is allocated to broadband real-time applications. In this book, our focus will be exclusively on FEC codes.

Forward error correction codes can be divided into two main classes of *Block Codes* and *Convolutional Codes.*

1.1.1 Block Codes

An (n, k) block code, with the same source and code alphabet, can be defined as a mapping from the k-dimensional extension of the source alphabet χ, i.e., χ^k into χ^n, $n > k$. The ratio $r = \frac{k}{n}$ is called the code rate. The alphabet used most often is the binary alphabet, i.e., $\chi = \{0, 1\}$. The set $\chi = \{0, 1\}$ with the modulo-2 addition (logical XOR) and modulo-2 multiplication (logical AND) forms a finite field of order 2, called a Galois Field and denoted as GF(2). A binary block code maps each k-bit source sequence $\mathbf{x} = (x_0, x_1, \ldots, x_{k-1})$ into an n-tuple $\mathbf{c} = (c_0, c_1, \ldots, c_{n-1})$ called a *codeword.* The code symbols c_j, $j = 0, \ldots, n-1$, are formed by combining the source symbols x_i, $i = 0, \ldots, k-1$ through different logical operations. Using only exclusive-OR (XOR), the resulting code will be a *linear code.* In a binary linear code, the sum of (for a non-binary code, any linear combination of) any two codewords is another codeword.

For linear codes, the encoding can be represented in terms of a matrix multiplication, i.e., $\mathbf{c} = \mathbf{x}.\mathbf{G}$, where $\mathbf{G}$ is a $k \times n$ matrix called the generator matrix.

A code is called systematic if each codeword consists of the k original information symbols plus $n-k$ redundant symbols (in the case of binary, the parity bits). In this case the generator matrix can be expressed as $\mathbf{G} = [\mathbf{I_k}|\mathbf{P}]$, where $\mathbf{P}$ is a $k \times (n-k)$ parity check matrix and $\mathbf{I_k}$ is a $k \times k$ identity matrix.

As an example consider the (7, 4) Hamming code with the generating matrix,

$$\mathbf{G} = \begin{bmatrix} 1 & 0 & 0 & 0 & 1 & 1 & 0 \\ 0 & 1 & 0 & 0 & 0 & 1 & 1 \\ 0 & 0 & 1 & 0 & 1 & 1 & 1 \\ 0 & 0 & 0 & 1 & 1 & 0 & 1 \end{bmatrix}. \tag{1.1}$$

Here, the first four columns form the systematic portion of the codeword by cloning the message bits, while the last three columns form the parity bits by each XOR-ing a different subset of the information bits.

Another interesting matrix defined for linear block codes is the parity check matrix $\mathbf{H}$ [34]. $\mathbf{H}$ is an $(n-k) \times n$ matrix such that $\mathbf{c}.\mathbf{H}^T = \mathbf{0}$ for any codeword $\mathbf{c}$, where $\mathbf{H}^T$ denotes the transpose of $\mathbf{H}$. Denoting a received vector by $\mathbf{y} = \mathbf{c}+\mathbf{e}$, where $\mathbf{c}$ is a transmitted codeword and $\mathbf{e}$ is an error pattern, i.e., an n-bit vector with 0's where the bits are received correctly and 1's in places where an error has occurred, we have $\mathbf{y}.\mathbf{H}^T = (\mathbf{c}+\mathbf{e}).\mathbf{H}^T = \mathbf{c}.\mathbf{H}^T + \mathbf{e}.\mathbf{H}^T = \mathbf{e}.\mathbf{H}^T$.

That is, the product of the received signal and the parity check matrix, called the *Syndrome* depends on the error pattern and not on the transmitted codeword. This property is used in algebraic decoding techniques. For a linear block code with $\mathbf{G} = [\mathbf{I_k}|\mathbf{P}]$, the parity check matrix is $\mathbf{H} = \left[\mathbf{P}^T|\mathbf{I_{n-k}}\right]$. For example, for the (7, 4) Hamming code the parity check matrix is given as,

$$\mathbf{H} = \begin{bmatrix} 1 & 0 & 1 & 1 & 1 & 0 & 0 \\ 1 & 1 & 1 & 0 & 0 & 1 & 0 \\ 0 & 1 & 1 & 1 & 0 & 0 & 1 \end{bmatrix}. \tag{1.2}$$

The error correcting and detecting capabilities of a code depend on the minimum distance between its codewords, i.e., the number of places two distinct codewords differ. A binary block code with minimum distance d_{min}, can detect any erroneous n-bit vector with up to $d_{min} - 1$ error bits and can correct error patterns with up to $t = \left\lfloor \frac{d_{min}-1}{2} \right\rfloor$ error bits, where $\lfloor z \rfloor$ denotes the largest integer no greater than z.

An important subclass of linear block codes consists of *cyclic* codes. In a cyclic code, any circular shift of a given codeword is another codeword. That is, if $\mathbf{c} = (c_0, c_1, \ldots, c_{n-1})$ is a codeword so is $\mathbf{c}^i = (c_i, c_{i+1}, \ldots, c_{n-1}, c_0, \cdots, c_{i-1})$. Due to this property, a code can be defined by a *generator polynomial* $g(X)$ and each codeword can be represented by a polynomial generated through multiplication of a polynomial representing the source data and $g(X)$. This facilitates the encoding and syndrome calculation through the use of linear feedback shift registers. It also results in various efficient algebraic decoding techniques.

1.1.2 Some Common Linear Block Codes

The first class of linear block codes used for error correction is the class of Hamming codes [24]. For any integer $m \geq 3$, we have a code with the following parameters:

Parameter	Value
Code Length	$n = 2^m - 1$
Number of information bits	$k = 2^m - m - 1$
Number of parity check bits	$n - k = m$
Minimum distance	$d_{min} = 3$
Error correcting capability	$t = 1$

The parity check matrix of a Hamming code is an $m \times (2^m - 1)$ matrix whose columns are all non-zero m-tuples [34].

The Bose-Chaudhuri-Hocquenghem (BCH) codes [25], [26], are a generalization of Hamming codes. They constitute a powerful class of binary block codes. For any positive integer $m \geq 3$, and $t \leq 2^m - 1$ there is a BCH code with the following parameters,

Parameter	Value
Code Length	$n = 2^m - 1$
Number of parity check bits	$n - k \leq mt$
Minimum distance	$d_{min} \geq 2t + 1$

By proper choice of m and t, one can select a BCH code suitable for a given channel condition. Generator polynomials for all binary BCH codes of length up to $2^{10} - 1$ are listed in [34].

The concept of binary codes can be extended to codes with non-binary alphabet. The most often used alphabets are the extensions of the binary alphabet. For any positive integer m, a Galois Field can be defined over the the alphabet consisting of all m-bit vectors. This filed is denoted as $GF(2^m)$. An (n, k) block code defined over $GF(2^m)$ consists of $(2^m)^k$ n-dimensional vectors whose elements are m-bit symbols belonging to $GF(2^m)$.

A widely used class of non-binary linear block codes is the class of *Reed Solomon* (RS) codes [27]. A t-error correcting RS code defined over $GF(2^m)$ has the following parameters,

Parameter	Value
Code Length	$n = 2^m - 1$
Number of parity check bits	$n - k = 2t$
Minimum distance	$d_{min} = 2t + 1$

Note that the code length is $2^m - 1$ symbols or $m \times (2^m - 1)$ bits. For example, for a byte oriented RS code where $m = 8$, the block length is 255 bytes or 2040 bits. The interest in RS codes is due to the fact that by correcting each symbol several bits in error are corrected. This makes them suitable for the situations where errors occur in bursts. Another interesting aspect of RS codes is that they are adaptable to different channel conditions and packet size constraints. The value of m can be selected to give a block length close to the desired packet size and error correcting capability can be adjusted by varying k. Furthermore, since the RS codes are systematic, they can be shortened by, conceptually, lengthening the information block by several 0 symbols and then encoding it. It is clear that these 0 symbols are deleted after encoding. In practice, it is sufficient to reset the encoder shift registers and then encode a block of $k - j$ symbols to get a shortened $(n - j, k - j)$ RS code. Since the number of parity symbols is still $n - k$ while the number of information symbols is reduced to $k - j$, the code is stronger.

An example of shortened Reed Solomon codes is the (204, 188) code for the *Digital Video Broadcasting* (DVB) standard used by digital TV broadcasters. This code is derived by shortening the (255, 239) code by 51 symbols. The length of the information field has been chosen 188 bytes to match the length of an MPEG (*Moving Picture Experts Group*) frame. While the original code corrects eight bytes in 255 bytes, the shortened code corrects 8 bytes in 204 bytes. In Digital Video Broadcasting-Return Channel via Satellite (DVB/RCS) standard [28], the (204, 188) RS code is used in the forward channel (from hub to remote station) while a shortened (73, 57) is used in the return channel (from remote terminal to hub)[2]. The choice of block length is dictated here by the desire to have a short ATM-like cell to ensure efficiency for interactive traffic[3].

1.1.3 Convolutional Codes

Encoders of block codes are one shot encoders. They take a k-symbol input block and encode it into an n-symbol codeword, where k and n are fixed and relatively long. *Trellis codes*, on the other hand, encode the data continuously and few bits at a time, thus, avoiding the fixed packet size. An (n, k, ν) binary convolutional encoder is a state machine with 2^ν states. At any given time t, it takes in k bits, outputs n bits and moves to a new state, k and n are relatively small. The generated output and the next state depend on the present state and the input,i.e.,

$$Y^t = f(S^t, X^t), \tag{1.3}$$

$$S^{t+1} = g(S^t, X^t), \tag{1.4}$$

where X^t and Y^t are the input and output at time t and S^t and S^{t+1} are the present and next states, respectively. Linear trellis codes, or, *convolutional* codes constitute a subclass of trellis codes where the input bits are passed through a linear shift register with ν k-bit stages and the output bits are formed by modulo-2 addition (XORing) of the most recent input symbol with different subsets of the outputs of the shift register cells. So each output bit can at most depend on $K = \nu + 1$ input symbols (the most recent symbol plus ν previous symbols), K is called the *constraint length* of the code. While it is possible to change the code rate $\frac{k}{n}$ by varying k, the common practice is to use *puncturing* in order to change the rate. Puncturing consists of deleting some of the bits at the output of the encoder. For example, rate 1/2 code can be changed into a rate 3/4 code by deleting every 4th output bit. Similarly, a rate 2/3 code can be formed

[2]Use of double binary Turbo codes is now in the DVB/RCS standard as an option.

[3]The cell consists of a regular 53-byte ATM cell plus 4 bytes of satellite specific data

by deleting two out of every six encoded bits. The punctured codes while being suboptimal compared to codes designed for a particular rate are popular due to the fact that a single encoder/decoder pair can be used for implementing different rates. This allows adapting the coding rate to the channel condition very easily. The convolutional code used most often is a rate 1/2 code with $\nu = 6$ [32]. In addition to rate 1/2, this code is used for encoding at rates 2/3, 3/4, 5/6 and 7/8 [28].

A convolutional encoder, like any finite state machine, can be represented using either a *state diagram,* or a *trellis* with 2^ν states. Each encoded sequence is represented as a path through the trellis. Figure 1.2 shows the encoder for a rate 1/2 binary code with constraint length $K = 3$ and Figure 1.3 shows its corresponding trellis. The bits on each branch represent the encoder output while the input bit is represented by the line style, i.e., solid line for zero and broken line for one.

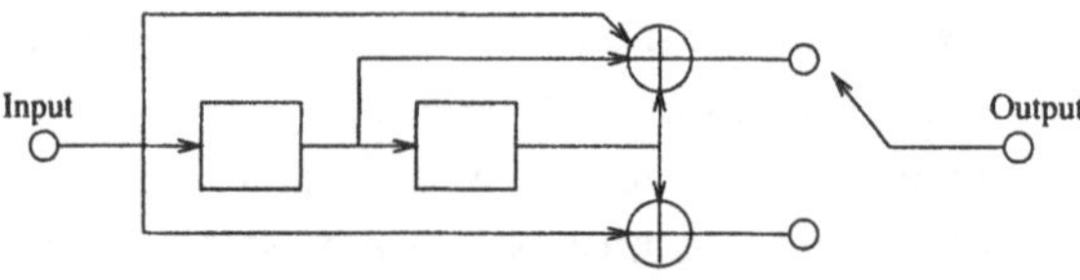

Figure 1.2. Block Diagram of a Convolutional Encoder

A convolutional code can be represented using *generator polynomials.* In general, there are $k \times n$ generator polynomials, i.e., one per each input-output pair. Each generator polynomial $g_i^{(j)}(D)$ is a degree ν polynomial where the coefficient D^p is 1 if the p-th shift of the j-th input takes part in forming of the i-th output. In the case of a rate $\frac{1}{n}$ convolutional code, there are only n generator polynomials. For the code of Figure 1.2, the two generator polynomials are $g_1(D) = 1 + D + D^2$ and $g_2(D) = 1 + D^2$.

An important parameter of a convolutional code is the minimum *free distance,* d_f [34]. Free distance is the minimum *Hamming distance* between two paths diverging at a given time instant and converging at a later time instant. Due to the linearity of convolutional codes, it suffices to consider paths in reference with the all zero path, i.e., to find the minimum weight of any path diverging and emerging to the all zero path. For the code of Figure 1.2, for example, $d_f = 5$. This corresponds to the input sequence 100 and output sequence 11, 10, 11 (see Figure 1.3).

For high $\frac{E_b}{N_0}$, using d_f only, i.e., considering the most probable error event, one can closely approximate the probability of error. However, for lower values of $\frac{E_b}{N_0}$, we need to take into consideration the effect of other error events and their multiplicity. For a binary convolutional code a close upper bound on the

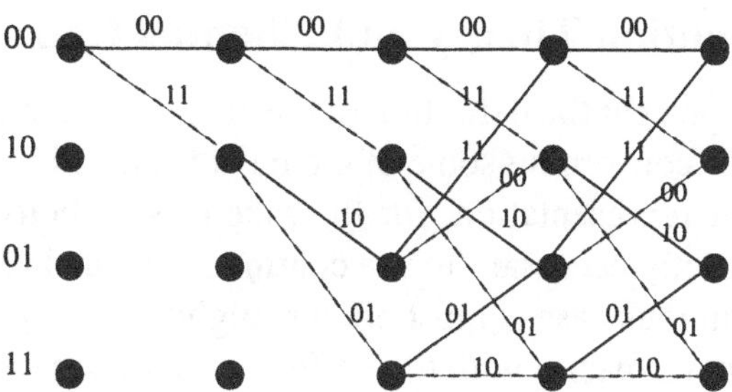

Figure 1.3. Trellis for the Convolutional Encoder of Figure 1.2

BER, in the case of soft decision decoding is given by [22],

$$P_b \leq Q\left(\sqrt{2rd_f\frac{E_b}{N_0}}\right) exp\left(rd_f\frac{E_b}{N_0}\right)\frac{\partial}{\partial I}T(X,\,Y)\Bigg|_{Y=1,\,X=exp(-rE_b/N_0)} \tag{1.5}$$

where r is the rate of the code and $T(X,\,Y)$ is the *augmented transfer function* of the code defined as,

$$T(X,\,Y) = \sum_{i=1}^{\infty}\sum_{d=d_f}^{\infty} a(d,\,i)X^dY^i, \tag{1.6}$$

where $a(d,\,i)$ is the number of paths with input weight i and output weight d.

For the case of hard decision decoding the BER is bounded as,

$$P_b \leq \frac{\partial}{\partial I}T(X,\,Y)\Bigg|_{Y=1,\,X=2\sqrt{p(1-p)}}, \tag{1.7}$$

where p is the error probability at the output of the demodulator. This is given, for example, for the BPSK (or QPSK) as,

$$p = Q\left(\sqrt{r\frac{2E_b}{N_0}}\right). \tag{1.8}$$

There are several algorithms for decoding of convolutional codes [34], however, the techniques used most often is the *Viterbi Algorithms* (VA) [29]. Given a channel output sequence, VA finds the sequence most likely to be the input to the encoder.

1.2. Information Theory and Channel Capacity

In this section, we present the definition of the channel capacity, its significance and a few examples concerning some of the most frequently encountered channel models. In our representation, for the sake of simplicity, we use discrete variables unless strictly necessary to use continuous valued variables. Generalization to the continuous case is most often straightforward [30]. Our goal is to compare the channel capacity with the performance of some of the well known conventional codes so that the reader can appreciate the improvement achieved using Turbo and Turbo-like codes.

A discrete memoryless channel is specified by an input source X taking values from an alphabet $\mathcal{X}$, an output Y taking values from an alphabet $\mathcal{Y}$ and a transition probability $Q = \{q(y|x) : x \in \mathcal{X}, y \in \mathcal{Y}\}$, where $q(y|x)$ is the conditional probability that the channel output is $Y = y$ when the input to the channel is $X = x$.

Definition 1.1: For a given probability distribution $P = \{p(x); x \in \mathcal{X}\}$ defined on X, the *entropy* of the source X is defined as,

$$H(X) = E[log(1/p(x))] = -\sum_x p(x)log[p(x)]. \tag{1.9}$$

$H(X)$ represents the amount of uncertainty about the random variable X, i.e., the average amount of information resolved by observation of specific realizations of X. Similarly,

$$H(Y) = E[log(1/q(y))] = -\sum_y q(y)log[q(y)] \tag{1.10}$$

is the uncertainty about the channel output Y. Here, $\{q(y)\}$ is the marginal probability distribution of Y given as,

$$q(y) = \sum_x p(x)q(y|x). \tag{1.11}$$

The conditional entropy of X given Y is given as,

$$H(X|Y) = -\sum_x \sum_y p(x)q(y|x)log[p(x|y)], \tag{1.12}$$

where,

$$p(x|y) = \frac{p(x)q(y|x)}{q(y)}. \tag{1.13}$$

$H(X|Y)$ is the uncertainty remaining about the channel input, X, after the observation of the channel output Y. The difference between $H(X)$ and $H(X|Y)$ is the average uncertainty resolved, i.e., the information gained, about

X from the observation of Y. For a communications engineer, this quantity represents the rate of information transfer through the channel,i.e.,

$$R = H(X) - H(X|Y). \tag{1.14}$$

The unit of R depends on the base of logarithm used in the definition of the entropies. If the base two is used, the unit is *bits per use.*

In this chapter and in the rest of the book, we always use base two for the logarithms so that the capacities calculated will be in units of *bits*, *bits per use* or *bits per second* depending on the context.

The quantity $H(X) - H(X|Y)$ can also be looked at as the average information provided about X from the observation of Y. In information theoretic parlance, this is called the average *mutual information*, $I(X;Y)$, i.e.,

$$I(X;Y) = H(X) - H(X|Y) = \sum_x \sum_y p(x)q(y|x) log \frac{p(x|y)}{p(x)}. \tag{1.15}$$

The rightmost equality is the result of combining Equations (1.9) and (1.12). It is easy to show that,

$$I(X;Y) = H(X) - H(X|Y) = H(Y) - H(Y|X), \tag{1.16}$$

that is, the information that Y provides about X is the same as that provide about Y by X. The *channel capacity* is then defined as the maximum transmission rate across the channel.

Definition 1.2: The *capacity* C of a discrete memoryless channel is defined as[30],

$$C = \max_P I(X;Y) = \max_P \sum_x \sum_y p(x)q(y|x) log \frac{q(y|x)}{\sum_x p(x)q(y|x)}, \tag{1.17}$$

where maximization is performed over all possible source distributions $P = \{p(x)\}$.

The importance of the channel capacity is due to the channel coding theorem and its converse that establish the channel capacity as the maximum rate of data transfer through a given channel. The channel coding theorem indicates that it is possible to find error control codes with rates arbitrarily close to C and with arbitrarily small probability of error. The converse to the channel coding theorem, on the other hand, says that the probability of error of any code with a rate $R > C$ is bounded away from zero [30].

The most basic discrete memoryless channel is a *Binary Symmetric Channel* (BSC). The input and the output of this channel take two values, usually denoted as *zero* and *one*. The probability of a *one* input being received, at the output of the channel, as a *zero* or vice versa is denoted as ε. The BSC models any binary communication system with an optimal symbol-by-symbol detection scheme assuming that the input bits are equally likely and the noise distribution is symmetric. This includes, for example, *Binary Phase Shift Keying* (BPSK)[4] with the *Maximum Likelihood* (ML) detection over *Additive White Gaussian Noise* (AWGN) channel. In this case, the crossover probability ε is the *Bit Error Rate* (BER) given as [20],

$$\varepsilon = Q\left(\sqrt{\frac{2E_b}{N_0}}\right). \tag{1.18}$$

For a BSC channel with crossover probability ε, the capacity is given by [30],

$$C = 1 + \varepsilon log\varepsilon + (1-\varepsilon)log(1-\varepsilon) = 1 - H_b(\varepsilon), \tag{1.19}$$

where $H_b(\varepsilon) = -\varepsilon log\varepsilon - (1-\varepsilon)log(1-\varepsilon)$ is the entropy of a binary source producing ones and zeros with probabilities ε and $1-\varepsilon$, respectively. The capacity given by Equation (1.19) is obtained by a source generating zeros and ones with the same probability. This should not come as a surprise since the assumptions made leading to use of ML detection include the assumption of equiprobable input bits.

For a channel with continuous input and continuous output, the capacity is defined as,

$$C = \max_P I(X;Y) = \max_P \int_x \int_y p(x)q(y|x)log\frac{q(y|x)}{q(y)}dxdy \tag{1.20}$$

where maximization is performed over all probability distributions satisfying a given constraint such as the average power constraint given as,

$$\int_x |x|^2 p(x)dx \leq S. \tag{1.21}$$

The channel model used most often is the AWGN channel. In an AWGN channel, independent identically distributed noise samples are added to the transmitted information symbols. The noise samples have a Gaussian distribution, i.e., the conditional density of the channel output y given the input x is given by,

$$q(y|x) = \frac{1}{\sqrt{2\pi}\sigma}e^{-(y-x)^2/2\sigma^2} \tag{1.22}$$

[4] Some complex quaternary modulation schemes such as Quaternary Phase Shift Keying (QPSK) can also be represented by BSC due to their separability into two binary modulation schemes.

Substituting Equation (1.22) into Equation (1.20) and maximizing with respect to **P**, the capacity of the AWGN channel is found to be [30],

$$C = Wlog\left(1 + \frac{S}{N}\right), \tag{1.23}$$

where W is the bandwidth occupied by the information bearing signal, S is the signal power and $N = WN_0$ is the Gaussian noise variance. Denoting the bit rate by R (in *bits per second*) and the energy per bit by E_b (in *joules*), we have $S = E_bR$. Substituting this into Equation (1.23) we get,

$$C = Wlog\left(1 + \frac{E_bR}{N_0W}\right) \tag{1.24}$$

The ratio $\eta = \frac{R}{W}$ is called the *spectral efficiency* or *spectral bit rate* measured in bits per second per Hertz.

From coding theorem, we know that in order to be able to communicate with arbitrarily low probability of error, the transmission rate should not exceed the channel capacity, i.e., $R < C$. Applying this constraint to Equation. (1.24), we get a lower bound on the $\frac{E_b}{N_0}$ required for a given spectral efficiency.

$$\frac{R}{W} < \frac{C}{W} = log\left(1 + \frac{R}{W}\cdot\frac{E_b}{N_0}\right) \tag{1.25}$$

Solving Equation. (1.25) for $\frac{E_b}{N_0}$, we get,

$$\frac{E_b}{N_0} > \frac{2^{\frac{R}{W}} - 1}{\frac{R}{W}} = \frac{2^{\eta} - 1}{\eta}. \tag{1.26}$$

Figure 1.4 shows the achievable spectral efficiency for different values of $\frac{E_b}{N_0}$. The points below the solid curve (the capacity curve) indicate the region where reliable communication is possible [31] while the points above the capacity curve represent the region where reliable communication is not possible. The most interesting aspect of the Shannon Theory, as expressed in the channel coding theorem, is that it not only gives, for any value of $\frac{E_b}{N_0}$, the range of the achievable rates, but also, indicates that one can approach the boundary between the achievable and unachievable rates as closely as desired.

An important point on the capacity curve is the point corresponding to $\frac{R}{W} = 0$, i.e., when there is no restriction on the bandwidth. Practically, it represents the situation where very low rate error control codes are used. For $\eta = \frac{R}{W} \to 0$, from Equation. (1.26) we get $\frac{E_b}{N_0} > ln2 = 0.69$ or, equivalently,

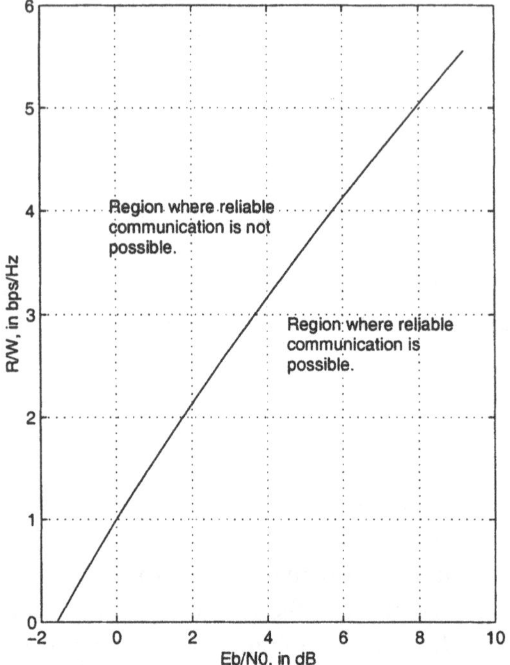

Figure 1.4. The Capacity of the AWGN Channel

$-1.6\ dB$. This means that, in an AWGN channel, if $\frac{E_b}{N_0} < -1.6\ dB$, then reliable communication is not possible no matter how many parity bits are added to the message bits

In digital communication systems, the input alphabet is finite while the output may be discrete or continuous. The former corresponds to hard decision decoding and the latter corresponds to soft decision decoding. In the case of hard decision decoding, a detector (demodulator) makes a tentative decision and provides it to the decoder. The decoder, then, tries to correct errors, possibly, made by the detector, hence the term *error correcting codes*. However, in the case of soft decision decoding, the channel output (the output of the *matched filter*) is passed directly to the decoder. The hard decision case can be modeled in terms of a channel with discrete input and discrete output and, therefore, the capacity in this case is given by Equation (1.17). For the case of soft decision decoding, the channel input is restricted to the constellation points of the modulation used, but there is no restriction on the channel output. Denoting the constellation points by $\{c_m : m = 1, ..., M - 1\}$, the channel capacity is defined as [31],

$$C = \max_{P} \sum_{m=0}^{M-1} p_m \int_{-\infty}^{\infty} q(y|c_m) log \frac{q(y|c_m)}{\sum_k p_k q(y|c_k)} \tag{1.27}$$

where p_m is the probability that the constellation point c_m be used and the maximization is performed with the constraint that the average transmitted power is,

$$S = \sum_{m=0}^{M-1} p_m c_m^2. \tag{1.28}$$

For a Gaussian channel $q(y|c_m)$ is given as,

$$q(y|c_m) = \frac{1}{\sqrt{2\pi}\sigma} e^{-(y-c_m)^2/2\sigma^2} \tag{1.29}$$

Here, for notational convenience, we have considered the scalar channel, i.e., a one dimensional modulation scheme such as M-ary PAM or M-ary ASK. Generalization to vector Gaussian-noise channel, e.g., for M-ary PSK or M-ary QAM, is straightforward and can be done by simple modifications to Equations (1.27)-(1.29). These modifications include replacing the single integral in Equation (1.27) by a double or multiple integral and modifying the norm and distance. A straightforward, though most often prohibitively complex, approach to the calculation of the capacity is to first fix the constellation, i.e., the c_m's and then maximize the Equation (1.27) by proper choice of **P**. However, it is usually reasonable to assume that the channel input probabilities are equal, i.e., $p_m = \frac{1}{M}$, $m = 0, \ldots, M-1$ [31]. Assuming that the inputs to the channel are equally probable, for a one dimensional constellation with equidistant points, Equation (1.27) can be written as,

$$C = \frac{1}{M} \sum_{m=0}^{M-1} \int_{-\infty}^{\infty} \frac{1}{\sqrt{2\pi}\sigma} e^{-(y-c_m)^2/2\sigma^2} log \frac{e^{-(y-c_m)^2/2\sigma^2}}{M^{-1} \sum_k e^{-(y-c_k)^2/2\sigma^2}} \tag{1.30}$$

Here the *Signal-to-Noise-Ratio* (SNR) is given by,

$$\frac{S}{N} = \frac{1}{M} \sum_{m=0}^{M-1} \frac{c_m^2}{\sigma^2}. \tag{1.31}$$

Figure 1.5 shows the capacity for one dimensional constellations with 2, 4, 8 and 16 points as well as the capacity of the Gaussian noise channel.

Figure 1.5 shows that the choice of modulation constellation depends on the transmitter's power. For example, it shows that for very low power, i.e., S/N

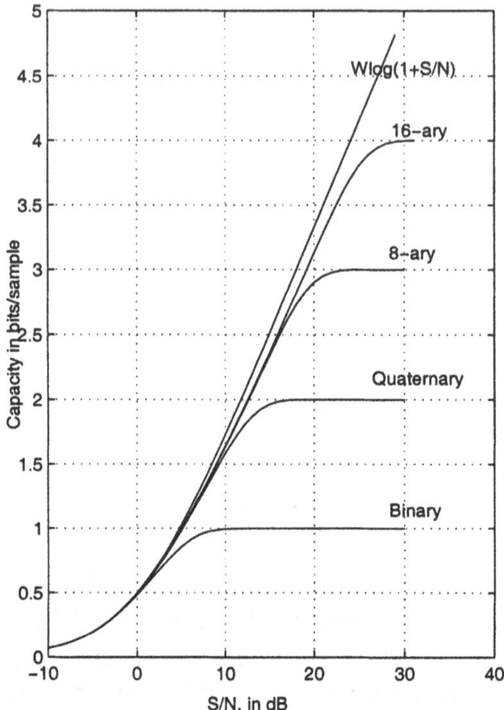

Figure 1.5. Capacity of Amplitude Modulation Schemes in AWGN Channel

below 3 dB binary modulation can realize most of the capacity of the AWGN channel and there is no point in using a modulation scheme with higher number of points. It is true that the bandwidth efficiency of the binary modulation is limited to 1 bit/sample, but using larger constellation does not allow crossing this barrier. The "extra" efficiency gained by using, say quaternary modulation, will be well compensated by the necessity of using lower rate error control coding scheme. The same way a four point modulation scheme is good up to around 10 dB of S/N and 8-ary constellation achieves most of the channel capacity up to 18 dB. In brief using a higher modulation scheme should be justified by the need for higher bandwidth efficiency and be backed by increased power.[5]

Figure 1.6 shows the capacity curves for some two dimensional modulation schemes such as MPSK (*M-ary Phase Shift Keying*) and QAM (*Quadrature Amplitude Modulation*) with different number of constellation points [31].

[5]In Chapter 4, we will see that using 8-PSK instead of QPSK we can increase the bandwidth efficiency beyond 2 bps/Hz and reduce the computational complexity while increasing the coding gain is almost impossible at low SNR.

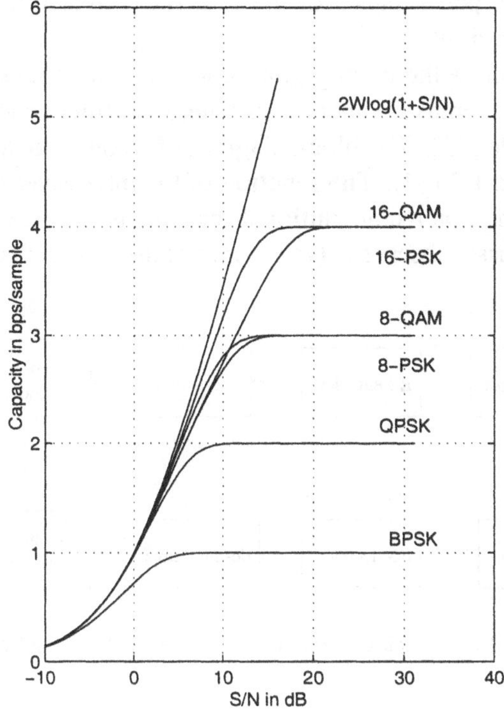

Figure 1.6. Capacity of 2-Dimensional Modulation Schemes in AWGN Channel

1.3. The Magic of Turbo Codes

In this section, we compare the performance of some of the conventional error control coding schemes with the channel capacity discussed in the previous section. By conventional codes, we mean the codes used, and still in use, in different digital communication systems, prior to the invention of Turbo codes. Then we contrast the performance of the Turbo and Turbo-like codes with the conventional codes as well as the channel capacity.

As a point of reference, we consider a digital communications system using BPSK (or QPSK) modulation with an error control code with a rate $r < 1$. For example, assume using the rate 1/2 convolutional code with constraint length $K = 7$ that was, until recently, the *de facto* industry standard. From [32], using this code with soft-decision Viterbi decoding, the $\frac{E_b}{N_0}$ required for achieving a BER of 10^{-5} is 4.4 dB. This is not such bad performance given the fact that uncoded QPSK requires an $\frac{E_b}{N_0}$ of 9.5 dB for the same BER, i.e., the convolutional code provides 5.1 dB of coding gain for only 3 dB increase in bandwidth requirement (it requires twice the bandwidth of the uncoded QPSK). Now, let's compare the performance of this code with the Shannon limit. Figure 1.5 (also from Figure 1.4) show that the required $\frac{E_b}{N_0}$ for *error free* transmission is zero dB. So, the performance of the rate 1/2 convolutional code is 4.4 dB above the Shannon limit.[6]

In order to improve the coding gain, convolutional codes are used in concatenated coding schemes as inner codes where the outer code is usually a Reed Solomon (RS) code [33]. The block diagram of a concatenated coding scheme is shown in Figure 1.7 [34]. The function of the interleaver between the *outer* RS encoder and the *inner* convolutional encoder is to spread the error bursts, i.e., to turn error bursts observed at the output of the Viterbi decoder into random errors.

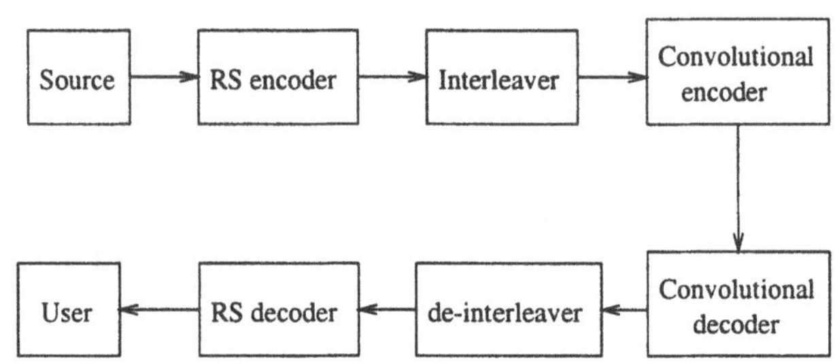

Figure 1.7. Block Diagram of a Concatenated Coding Scheme

The interleaver is only effective in reducing the BER if it is long enough to span several RS codewords. In such a case, it divides the error bursts amongst the different RS codewords in such a way that the number of erroneous symbols in each codewords (frame) is within the error correcting capability of the RS code. In cases where small packets of data are transmitted, e.g., DVB/RCS where the cells are 204 bytes in the forward link and 57 bytes on the return link, the use of interleaving is impossible or at least limited. At this point, we assume long blocks of data and ideal interleaving in order to assess the best performance that the above concatenated scheme can provide and compare with what is achieved using turbo codes with large interleaver. Later, we compare the two schemes for the case of short packets.

The performance of the concatenated coding system depends on the rate of the inner and outer code as well as the size of the interleaver. The best performance reported for a BER of 10^{-5} [34] is 2.81 dB for a (255, 239) RS code defined over Galois Field GF(2^8). In this field each symbol is 8-bits long and therefore the total length of the codeword is $255 \times 8 = 2040$ bits. For this scheme the overall $\frac{R}{W}$ is $\frac{239}{255}$ and from Equation (1.26) we have the required $\frac{E_b}{N_0}$ for error free transmission is -0.1 dB. So, this point is 2.91 dB away from

[6] In modern broadband communication systems usually a BER in the range of 10^{-12} to 10^{-10} is usually considered as "error free". However, we use the figure 10^{-5} since it has been widely used in the coding literature as a reference point.

the Shannon limit. Another arrangement using the same inner convolutional code with a (255, 191) RS code gives a BER of 10^{-5} [34] at 2.5 dB. However, for this arrangement, the Shannon limit is -0.42 dB (corresponding to a rate of 191/255) and therefore the deviation from the theoretical limit is almost the same.

Now let's turn to the Turbo codes and compare their performance with the information theoretical limit. In [6] parallel concatenation of two rate 1/2 systematic recursive convolutional codes are used with an iterative decoding algorithm involving the exchange of information between two decoders. The innate rate of the code is 1/3 as the systematic part is sent plus parities from each of the two encoders. However, the code rate could be reduced through puncturing. The resulting 1/2 rate turbo code [6] gave a BER of 10^{-5} at $\frac{E_b}{N_0} =$ 0.7 dB with a data block (interleaver) size of $N = 65536$ bits and after 18 iterations. This is only 0.7 dB off the Shannon limit.

Later Nickl et al. [35] devised a turbo code based on the concatenation of two (1023, 1013) Hamming codes operating within 0.27 dB of the channel capacity limit of a *binary input* AWGN channel with only 10 iterations. It is instructive to compare the performance of this code with the uncoded case since this code achieves a BER of 10^{-5} at $\frac{E_b}{N_0} \approx 5.6$ dB with a coding rate of 0.9805, i.e., it demonstrates 3.9 dB coding gain at almost no bandwidth penalty.

It is important to note the difference between achieving the channel capacity limit for an AWGN channel with arbitrary input constellation and one with input alphabet constrained to a given constellation. For example, while the code of [35] is closer than the code presented by Berrou et al. [6] to the channel capacity limit of an AWGN channel with binary input, it is farther from the channel capacity limit of an AWGN channel without input constraint. The reason is that in the former case, i.e., at a rate of 1/2, the binary input constellation realizes almost all the capacity of the AWGN channel, while in the latter case, the code is performing at full rate where the difference in the general channel capacity limit and the binary input capacity limit is considerable[7]. As discussed earlier, this should be considered as an indication that a move to a larger constellation is needed. In satellite and wireless communications terminology, this is a move from a *power limited* regime to a *bandwidth limited* regime, i.e., taking the advantage of the excess power to increase bandwidth efficiency.

The above results are achieved with very long data blocks. In many applications, using such long block sizes result in transmission inefficiency, particularly for interactive services. This limitation while affecting the performance of the turbo coding schemes, has even more drastic effect on the conventional coding schemes. For example, while the concatenation of an 8-error correcting RS

[7] At rate 0.5, the AWGN limit is 0 dB and the binary input limit is 0.19 dB while at r=0.9805, these values are 1.69 dB and 5.33 dB, respectively.

code and a rate 1/2, K=7 convolutional code is supposed to give capable of providing a BER of 10^{-6} at $\frac{E_b}{N_0} = 2.9\,dB$ with arbitrarily long interleaver [34] between the outer and inner codes, the (73, 57) suggested in DVB/RCS [28] standard with rate 1/2 convolutional code (overall rate of 0.39) and no interleaving achieves a BER of 10^{-6} at $\frac{E_b}{N_0} = 4.3\,dB$. The same BER can be obtained using the DVB/RCS turbo code at $\frac{E_b}{N_0} = 2.3\,dB$ at the rate of 1/2 with only four iterations [36]. Note that 2 dB improvement is achieved with better spectrum efficiency and at reasonably low complexity.

1.4. Outline of the Book

In this chapter, we presented some of the basic concepts of information theory and coding theory that we consider useful for appreciating the material covered in other chapters. Furthermore, we discussed the capability of Turbo codes in approaching the theoretical limits.

In Chapter 2, basic turbo coding principles including the concatenation of simple codes and interleaving to get powerful codes as well as the iterative decoding techniques will be presented. In Chapter 2, we will also give examples of binary turbo coding schemes including the turbo codes suggested for 3GPP wireless standard.

In Chapter 3, we extend the treatment of turbo codes from binary to non-binary codes. In particular, we will discuss in detail the transmission scheme consisting of a double-binary turbo code with QPSK modulation suggested in the DVB-RCS (*Digital Video Broadcasting-Return Channel via Satellite*) standard [28].

In Chapter 4, in order to overcome the 2 bps/Hz constraint imposed by the use of QPSK (*Quadrature Phase Shift Keying*) modulation, we present the use of a triple-binary turbo code with 8PSK (*8-ary Phase Shift Keying*) modulation.

In Chapter 5, application of turbo decoding principle to the concatenated codes consisting of block codes is discussed. In this chapter, different encoder configurations for *Block Turbo Codes* (BTCs) will be presented. Then various decoding algorithms for BTCs including *Maximum a posteriori Probability* (MAP) and Chase Algorithm will be discussed.

In Chapter 6, we discuss the *Reed Muller* (RM) codes and their trellis structure. Then, turbo codes with RM code as their constituent codes will be introduced and the trellis based iterative decoding algorithm for them will be presented.

In Chapter 7, The performance of different block turbo coding schemes will be reviewed. In this chapter, we will also present the application of BTC in several wireless and satellite communication systems and commercially available products.

In Chapter 8, we will discuss some implementation issues including the fixed point implementation and SNR mismatch. We will also discuss the implementation of Turbo codes using FPGA (*Field Programmable Gate Array*), ASIC (*Application Specific Integrated Circuit*) and general purpose DSPs (*Digital Signal Processings*).

In Chapter 9, we will discuss the related class of *Low Density Parity Check* (LDPC) codes.

Chapter 2

TURBO DECODING PRINCIPLES

We consider a one-way communication system, where the transmission is strictly in the forward direction, from the transmitter to the receiver. In contrast to a two-way system that can use ARQ with error detection and retransmission, the error control strategy for a one-way system must be FEC, which automatically corrects errors detected at the receiver. The class of FEC codes includes block codes, convolutional codes, as well as concatenated codes that are built using block and/or convolutional codes. Turbo codes and LDPC codes are the newest members of the family of FEC codes.

In this chapter, after a brief review of basic ideas behind turbo codes, as well as LDPC codes that are going to be introduced in more detail in Chapter 9, turbo decoding principle will be presented. We will then discuss some of the issues such as the choice of the constituent codes, the interleaving, trellis termination and puncturing. Finally, we give some examples of turbo codes including the one in 3GPP standard.

2.1. Turbo Codes and LDPC codes

Concatenated codes were first proposed by Forney [46] as a means for achieving high coding gain (without the complexity of long codes) by combining two or more relatively simple component codes. A serial concatenation of codes is often used in power limited channels such as deep space and satellite communication applications. The most popular of these schemes is a serial concatenation of an outer Reed-Solomon code with an inner convolutional code [33]. Product codes introduced by Elias in 1954 [41], known for their simultaneous burst- and random-error correcting capability, can serve a similar purpose. Conventional algorithms for decoding concatenated and products codes, however, gave rather poor results because they used hard decision decoding. In 1992, John Lodge et al. proposed a solution with good performance based on iterative decoding using soft-input/soft-output decoders [4]. In 1993, Berrou [6] introduced a coding scheme consisting of two parallel recursive systematic convolutional encoders separated by an interleaver and using an iterative *A Posteriori Probability* (APP) decoder. The scheme, called Turbo coding, achieved an exceptionally low BER at a SNR very close to Shannon's theoretical limit.

The *Maximum a posteriori Probability* (MAP) algorithm was applied to the problem of symbol-by-symbol detection of coded sequences by Bahl, Cocke, Jelinik and Raviv in 1974 [57]. The resulting algorithm, called the BCJR algorithm became popular in the research community because of the introduction of turbo codes in recent years.

Motivated by the introduction of turbo codes, also called *Parallel Concatenated Convolutional Codes* (PCCCs), *Serial Concatenated Convolutional Codes* (SCCCs) [59] and *Hybrid Concatenated Convolutional Codes* (HCCCs) were later constructed providing similar, and sometimes even better, coding gains compared to PCCCs [58]. At high signal-to-noise ratios, because of a superior distance profile, SCCC and HCCC can outperform PCCC. In addition to binary convolutional codes, non-binary convolutional codes and block codes such as Hamming codes, RM codes and RS codes, can also be used as the constituent code in the concatenation scheme.

Another important class of linear block codes, Gallager's LDPC codes [162], has received considerable attention prompted by the invention of turbo codes and unprecedented increase in the computing power of digital circuitry. With iterative *message-passing* decoding algorithms, extended LDPC coding techniques have been shown to exhibit performance comparable to, and sometimes even better than, the original turbo codes. The extension of Gallager's technique of "density evolution" has provided a breakthrough in the analysis of such iterative message-passing decoders, as well as the basis for a practical design method for powerful LDPC codes on a large class of channels. Beginning with a specified class of bipartite graphs and the corresponding ensemble of LDPC codes, the technique determines a threshold value that can be translated into a minimum signal-to-noise-ratio [178], above which the message-passing decoder will yield asymptotically good performance for most codes in the associated LDPC code family. For optimized graph structures, the resulting thresholds have been shown, in some cases, to be extremely close to those corresponding to the Shannon capacity and simulations with large block lengths have confirmed good code performance essentially at the threshold [178].

The decoding principle used for turbo codes and LDPC codes is now widely recognized as a very general and powerful concept in communication theory, with applications that go beyond the practical decoding of these codes. The *turbo principle* [7], describes the fundamental strategy underlying the success of turbo decoding, namely, the iterative exchange of soft information between different blocks in a communications receiver in order to improve overall system performance.

2.2. Iterative Decoding Principle

An iterative turbo decoder consists of two component decoders concatenated serially via an interleaver, identical to the one in the encoder. SISO (*Soft Input/Soft Output*) algorithms are well suited for iterative decoding because they accept *a priori* information at their input and produce *a posteriori* information at their output. In turbo decoding, trellis based decoding algorithms are used. These are recursive methods suitable for the estimation of the state sequence of a discrete-time finite-state Markov process observed in memoryless noise. With reference to decoding of noisy coded sequences, the MAP algorithm is used to estimate the most likely information bit to have been transmitted in a coded sequence. Here, we only discuss the iterative decoding of two-dimensional turbo codes. The extension to the case of multidimensional concatenated codes is straightforward.

2.2.1 BCJR Algorithm

The Bahl, Cocke, Jelinek, and Raviv (BCJR) algorithm [57], also known as the forward-backward or the *a posteriori* probability algorithm, or Maximum *a posteriori* algorithm, is the core component in many iterative detection and decoding schemes. BCJR algorithm is optimal for estimating the states or the outputs of a Markov process observed in white noise. It produces the sequence of *A Posteriori Probabilities* (APP), $P(u_k = i|observation)$, $i = 0,\ 1$, where $P(u_k = i|observation)$ is the APP of the data bit u_k given all the received sequence. The numerical representation of probabilities, non-linear functions and mixed multiplications and additions of these values perhaps make this algorithm too difficult to implement. As a result, different derivatives of this algorithm such as Log-MAP and Max-Log-MAP algorithm have been used in the decoding of turbo codes. Another approach is to use SOVA.

2.2.2 Tools for Iterative Decoding of Turbo Codes

We first describe the turbo principle restricted to the case of binary and *Recursive Systematic Convolutional* (RSC) codes and later, in Chapter 3 and 4, extend it to the case of non-binary and *Circular Recursive Systematic Convolutional* (CRSC) codes . The decoding principle for block turbo codes will be introduced in Chapter 5.

2.2.2.1 Log-likelihood Algebra. The log-likelihood ratio of a binary random variable u_k, $L(u_k)$, is defined as

$$L(u_k) = \ln \frac{P(u_k = +1)}{P(u_k = -1)} \tag{2.1}$$

where u_k is the information bit at time k. It is in GF(2) with the elements $\{+1, -1\}$, and +1 is the "null" element under the $\oplus$ addition.

Since

$$P(u_k = +1) = 1 - P(u_k = -1) \tag{2.2}$$

and

$$L(u_k) = \ln \frac{P(u_k = +1)}{1 - P(u_k = +1)} \tag{2.3}$$

then

$$e^{L(u_k)} = \frac{P(u_k = +1)}{1 - P(u_k = +1)} \tag{2.4}$$

Hence,

$$\begin{aligned} P(u_k = +1) &= \frac{e^{L(u_k)}}{1 + e^{L(u_k)}} \\ &= \frac{1}{1 + e^{-L(u_k)}} \\ &= (\frac{e^{-L(u_k)/2}}{1 + e^{-L(u_k)}}) \cdot e^{L(u_k)/2} \end{aligned} \tag{2.5}$$

and

$$\begin{aligned} P(u_k = -1) &= 1 - \frac{1}{1 + e^{-L(u_k)}} \\ &= \frac{e^{-L(u_k)}}{1 + e^{-L(u_k)}} \\ &= (\frac{e^{-L(u_k)/2}}{1 + e^{-L(u_k)}}) \cdot e^{-L(u_k)/2} \end{aligned} \tag{2.6}$$

Equations (2.5) and (2.6) can be represented as

$$\begin{aligned} P(u_k = \pm 1) &= \frac{e^{\pm L(u_k)}}{1 + e^{\pm L(u_k)}} \\ &= (\frac{e^{-L(u_k)/2}}{1 + e^{-L(u_k)}}) \cdot e^{u_k \cdot L(u_k)/2} \\ &= A_k \cdot e^{u_k \cdot L(u_k)/2} \end{aligned} \tag{2.7}$$

where $A_k = (\frac{e^{-L(u_k)/2}}{1+e^{-L(u_k)}})$ is a common factor.

If the binary random variable u_k is conditioned on a different random variable or vector y_k, then we have a conditioned log-likelihood ratio $L(u_k|y_k)$ with

$$L(u_k|y_k) = \ln \frac{P(u_k = +1|y_k)}{P(u_k = -1|y_k)}$$

$$
\begin{aligned}
&= \ln \frac{p(y_k|u_k=+1)\cdot P(u_k=+1)}{p(y_k|u_k=-1)\cdot P(u_k=-1)} \\
&= \ln \frac{p(y_k|u_k=+1)}{p(y_k|u_k=-1)} + \ln \frac{P(u_k=+1)}{P(u_k=-1)} \\
&= L(y_k|u_k) + L(u_k) \qquad (2.8)
\end{aligned}
$$

2.2.2.2 Soft Channel Outputs. After transmission over a channel with a fading factor a and additive Gaussian noise,

$$
\begin{aligned}
L(u_k|y_k) &= \ln \frac{p(y_k|u_k=+1)\cdot P(u_k=+1)}{p(y_k|u_k=-1)\cdot P(u_k=-1)} \\
&= \ln \frac{exp(-\frac{E_s}{N_0}(y_k-a)^2)}{exp(-\frac{E_s}{N_0}(y_k+a)^2)} + \ln \frac{P(u_k=+1)}{P(u_k=-1)} \\
&= \ln\{exp[-\frac{E_s}{N_0}(y_k^2-2\cdot a\cdot y_k+a^2-y_k^2-2\cdot a\cdot y_k-a^2)]\} \\
&\quad +L(u_k) \\
&= 4\cdot a\cdot \frac{E_s}{N_0}\cdot y_k + L(u_k) \\
&= L_c\cdot y_k + L(u_k) \qquad (2.9)
\end{aligned}
$$

where $L_ç = 4 \cdot a \cdot \frac{E_s}{N_0}$. For a fading channel, a denotes the fading amplitude whereas for a Gaussian channel, we set $a = 1$. For a *Binary Symmetric Channel* (BSC), we have the same relationship where L_c is the log-likelihood ratio of the crossover probabilities p and $1-p$, *i.e.*, $L_c = log((1-p)/p)$. L_c is called the reliability value of the channel [7].

Since

$$
P(y_k) = p(y_k|u_k=+1)\cdot P(u_k=+1) + p(y_k|u_k=-1)\cdot P(u_k=-1) \qquad (2.10)
$$

$$
p(y_k|u_k=-1) = \frac{P(y_k)}{P(u_k=-1)} - \frac{p(y_k|u_k=+1)\cdot P(u_k=+1)}{P(u_k=-1)} \qquad (2.11)
$$

and

$$
\begin{aligned}
L(y_k|u_k) &= \ln \frac{p(y_k|u_k=+1)}{\frac{P(y_k)}{P(u_k=-1)} - \frac{p(y_k|u_k=+1)\cdot P(u_k=+1)}{P(u_k=-1)}} \\
&= L_c\cdot y_k \qquad (2.12)
\end{aligned}
$$

then

$$
e^{L_c\cdot y_k} = \frac{p(y_k|u_k=+1)}{\frac{P(y_k)}{P(u_k=-1)} - \frac{p(y_k|u_k=+1)\cdot P(u_k=+1)}{P(u_k=-1)}} \qquad (2.13)
$$

$$p(y_k|u_k=+1) = (\frac{P(y_k)}{P(u_k=-1)} - \frac{p(y_k|u_k=+1)\cdot P(u_k=+1)}{P(u_k=-1)})\cdot e^{L_c\cdot y_k} \tag{2.14}$$

Hence

$$p(y_k|u_k=+1) = \frac{\frac{P(y_k)}{P(u_k=-1)}\cdot e^{L_c\cdot y_k}}{1+\frac{P(u_k=+1)}{P(u_k=-1)}\cdot e^{L_c\cdot y_k}} \tag{2.15}$$

substitute $P(u=+1)$ and $P(u=-1)$ as in Equations (2.5) and (2.6):

$$\begin{aligned} p(y_k|u_k=+1) &= \frac{\frac{P(y_k)\cdot(1+e^{L_c\cdot y_k})}{e^{-L(u_k)}}\cdot e^{L_c\cdot y_k}}{1+e^{L(u_k)}\cdot e^{L_c\cdot y_k}} \\ &= \frac{\frac{P(y_k)\cdot(1+e^{L_c\cdot y_k})}{e^{-L(u_k)}}\cdot e^{L_c\cdot y_k}\cdot e^{-(L(u_k)+L_c\cdot y_k)}}{e^{-(L(u_k)+L_c\cdot y_k)}+1} \\ &= \frac{P(y_k)\cdot(1+e^{-L(u_k)})}{1+e^{-(L(u_k)+L_c\cdot y_k)}} \end{aligned} \tag{2.16}$$

similarly

$$p(y_k|u_k=+1) = \frac{P(y_k)}{P(u_k=+1)} - \frac{p(y_k|u_k=-1)\cdot P(u_k=-1)}{P(u_k=+1)} \tag{2.17}$$

$$\begin{aligned} L(y_k|u_k) &= \ln\frac{\frac{P(y_k)}{P(u_k=+1)} - \frac{p(y_k|u_k=-1)\cdot P(u_k=-1)}{P(u_k=+1)}}{p(y_k|u_k=-1)} \\ &= L_c\cdot y_k \end{aligned} \tag{2.18}$$

$$e^{-L_c\cdot y_k} = \frac{p(y_k|u_k=-1)}{\frac{P(y_k)}{P(u_k=+1)} - \frac{p(y_k|u_k=-1)\cdot P(u_k=-1)}{P(u_k=+1)}} \tag{2.19}$$

$$\begin{aligned} p(y_k|u_k=-1) &= \frac{\frac{P(y_k)}{P(u_k=+1)}\cdot e^{-L_c\cdot y_k}}{1+\frac{P(u_k=-1)}{P(u_k=+1)}\cdot e^{-L_c\cdot y_k}} \\ &= \frac{P(y_k)\cdot(1+e^{-L(u_k)})\cdot e^{-L_c\cdot y_k}}{1+e^{-(L(u_k)+L_c\cdot y_k)}} \end{aligned} \tag{2.20}$$

Hence

$$\begin{aligned} p(y_k|u_k=\pm1) &= (\frac{P(y_k)\cdot(1+e^{-L(u_k)})\cdot e^{-L_c\cdot y_k/2}}{1+e^{-(L(u_k)+L_c\cdot y_k)}})\cdot e^{u_k\cdot L_c\cdot y_k/2} \\ &= B_k\cdot e^{u_k\cdot L_c\cdot y_k/2} \end{aligned} \tag{2.21}$$

where $B_k = (\frac{P(y_k)\cdot(1+e^{-L(u_k)})\cdot e^{-L_c\cdot y_k/2}}{1+e^{-(L(u_k)+L_c\cdot y_k)}})$ is the common factor.

2.2.2.3 Principle of the Iterative Decoding Algorithm. Assume that we have a "soft-in/soft-out" decoder available as shown in Figure 2.1 [7] for decoding of the component codes.

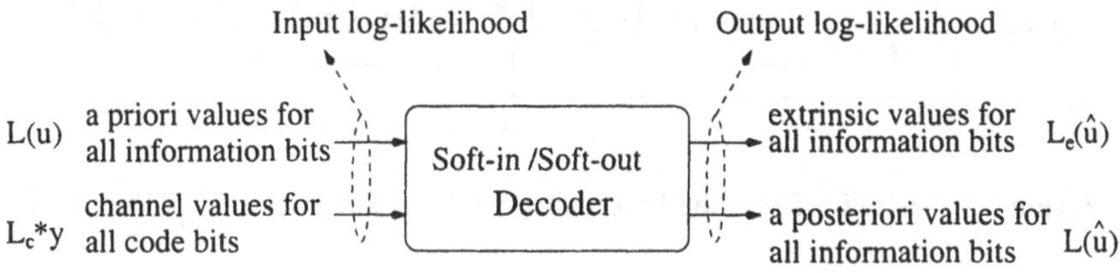

Figure 2.1. "Soft-in/Soft-out" Decoder

The output of the "symbol-by-symbol" *Maximum a posteriori Probability* (MAP) decoder is defined as the *a posteriori* log-likelihood ratio, that is, the logarithm of the ratio of the probabilities of a given bit being "+1" or "-1' given the observation y.

$$L(\hat{u}) = L(u|y) = ln\frac{P(u=+1|y)}{P(u=-1|y)} \tag{2.22}$$

Such a decoder uses *a priori* values $L(u)$ for all information bits u, if available, and the channel values $L_c \cdot y$ for all coded bits. It also delivers soft outputs $L(\hat{u})$ on all information bits and an extrinsic information $L_e(\hat{u})$, which contains the soft output information from all the other coded bits in the code sequence and is not influenced by the $L(u)$ and $L_c \cdot y$ values of the current bit. For systematic codes, the soft output for the information bit u will be represented as the sum of three terms

$$L(\hat{u}) = L_c \cdot y + L(u) + L_e(\hat{u}) \tag{2.23}$$

This means that we have three independent estimates for the log-likelihood ratio of the information bits: the channel values $L_c \cdot y$, the *a priori* values $L(u)$ and the values $L_e(\hat{u})$ by a third independent estimator utilizing the code constraint. The whole procedure of iterative decoding with two "Soft-in/Soft-out" decoders is shown in Figure 2.2.

In the first iteration of the iterative decoding algorithm, Decoder 1 computes the extrinsic information

$$L_e^1(\hat{u}) = L^1(\hat{u}) - [L_c \cdot y + L(u)] \tag{2.24}$$

We assume equally likely information bits: thus we initialize $L(u) = 0$ for the first iteration. This extrinsic information from the first decoder, is passed to the Decoder 2, which uses $L_e^1(\hat{u})$ as the *a priori* value in place of $L(u)$ to compute $L^2(\hat{u})$. Hence, the extrinsic information value computed by Decoder

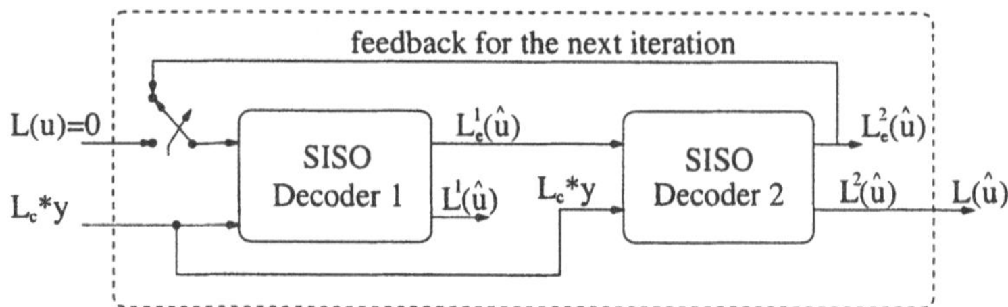

Figure 2.2. Iterative Decoding Procedure with Two "Soft-in/Soft-out" Decoders

2 is

$$L_e^2(\hat{u}) = L^2(\hat{u}) - [L_c \cdot y + L_e^1(\hat{u})] \tag{2.25}$$

Then, Decoder 1 will use the extrinsic information values $L_e^2(\hat{u})$ as *a priori* information in the second iteration. The computation is repeated in each iteration.

The iterative process is usually terminated after a predetermined number of iterations, when the soft-output value $L^2(\hat{u})$ stabilizes and changes little between successive iterations. In the final iteration, Decoder 2 combines both extrinsic information values in computing the soft-output values

$$L^2(\hat{u}) = L_c \cdot y + L_e^1(\hat{u}) + L_e^2(\hat{u}) \tag{2.26}$$

2.2.3 Optimal and Suboptimal Algorithms

The Maximum Likelihood Algorithms such as Viterbi Algorithm, find the most probable *information sequence* that was transmitted, while the MAP algorithm finds the most probable *information bit* to have been transmitted given the coded sequence. The information bits returned by the MAP algorithm need not form a connected path through the trellis.

For estimating the states or the outputs of a Markov process, the symbol-by-symbol MAP algorithm is optimal. However, MAP algorithm is not practicable for implementation due to the numerical representation of probabilities, non-linear functions and lot of multiplications and additions. Log-MAP algorithm avoids the approximations in the Max-Log-MAP algorithm and hence is equivalent to the true MAP but without its major disadvantages. MAP like algorithms, SOVA and the Max-Log-MAP algorithm, are both suboptimal at low signal-to-noise ratios. The relationship between these algorithms is illustrated in Figure 2.3.

2.2.3.1 MAP algorithm. The trellis of a binary feedback convolutional encoder has the structure shown in Figure 2.4.

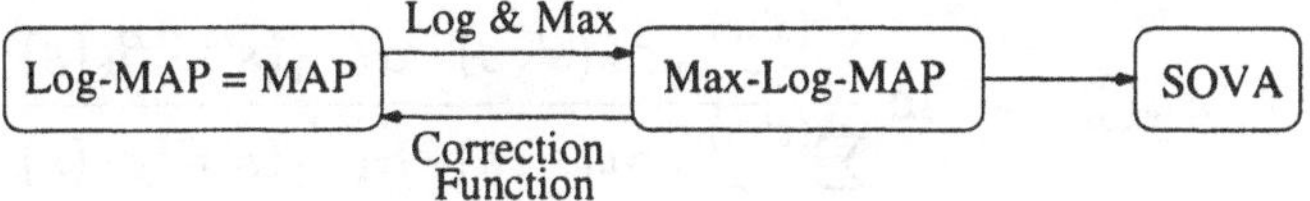

Figure 2.3. Relationship between MAP, Log-MAP, Max-Log-MAP and SOVA

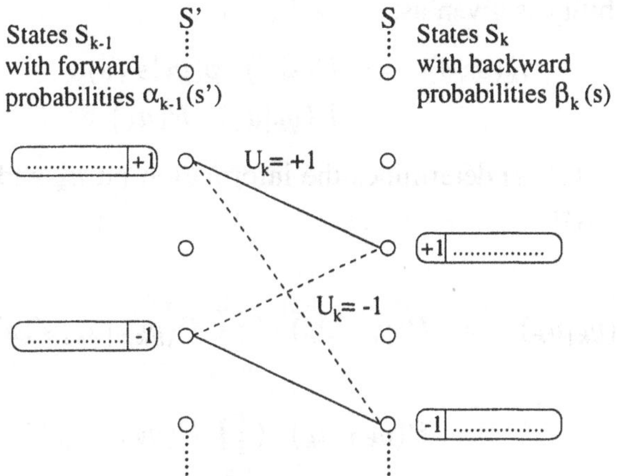

Figure 2.4. Trellis Structure of Systematic Convolutional Codes with Feedback Encoders

From above, define the log-likelihood ratio as:

$$L(\hat{u}_k) = \ln \frac{P(u_k = +1|y)}{P(u_k = -1|y)} = \ln \frac{\sum_{u_k=+1}^{(s',s)} P(s', s, y)}{\sum_{u_k=-1}^{(s',s)} P(s', s, y)} \tag{2.27}$$

where

$$\begin{aligned} P(s', s, y) &= P(s', y_{j<k}) \cdot P(s, y_k|s') \cdot P(y_{j>k}|s) \\ &= \underbrace{P(s', y_{j<k})}_{} \cdot \underbrace{P(s|s') \cdot P(y_k|s', s)}_{} \cdot \underbrace{P(y_{j>k}|s)}_{} \\ &= \alpha_{k-1}(s') \cdot \qquad \gamma_k(s', s) \cdot \qquad \beta_k(s) \end{aligned} \tag{2.28}$$

Here $y_{j<k}$ denotes the sequence of received symbols y_j from the beginning of the trellis up to time $k-1$ and $y_{j>k}$ is the corresponding sequence from time $k+1$ up to the end of the trellis. The forward recursion and backward recursion of the MAP algorithm yield

$$\alpha_k(s) = \sum_{(s',s)} \gamma_k(s', s) \cdot \alpha_{k-1}(s') \tag{2.29}$$

$$\beta_{k-1}(s') = \sum_{(s',s)} \gamma_k(s', s) \cdot \beta_k(s) \tag{2.30}$$

$$L(\hat{u}_k) = \ln \frac{\sum_{u_k=+1}^{(s',s)} \gamma_k(s',s) \cdot \alpha_{k-1}(s') \cdot \beta_k(s)}{\sum_{u_k=-1}^{(s',s)} \gamma_k(s',s) \cdot \alpha_{k-1}(s') \cdot \beta_k(s)} \quad (2.31)$$

Whenever, there is a transition from s' to s, $P(s|s') = P(u_k)$, where u_k is the information bit corresponding to the transition from s' to s and the branch transition probability is given as,

$$\begin{aligned} \gamma_k(s',s) &= P(s|s') \cdot p(y_k|s',s) \\ &= P(y_k|u_k) \cdot P(u_k) \end{aligned} \quad (2.32)$$

The index pair (s', s) determines the information bit u_k and the coded bits $x_{k,v}$, for $v = 2, ..., n$.

where

$$\begin{aligned} P(y_k|u_k) &= P(y_{k,1}|u_k) \cdot (\prod_{v=2}^{n} P(y_{k,v}|u_k, s', s)) \\ &= P(y_{k,1}|u_k) \cdot (\prod_{v=2}^{n} P(y_{k,v}|x_{k,v}) \end{aligned} \quad (2.33)$$

is the independent joint probabilities of the received symbols and

$$P(u_k) = P(u_k = \pm 1) = A_k e^{u_k L(u_k)/2} \quad (2.34)$$

From Equation (2.21), we have,

$$\begin{aligned} P(y_k|u_k) &= P(y_{k,1}|u_k) \cdot (\prod_{v=2}^{n} P(y_{k,v}|x_{k,v}) \\ &= B_k \cdot exp(\frac{1}{2} L_c \cdot y_{k,1} \cdot u_k) \cdot (\prod_{v=2}^{n} exp(\frac{1}{2} L_c \cdot y_{k,v} \cdot x_{k,v}) \\ &= B_k \cdot exp(\frac{1}{2} L_c \cdot y_{k,1} \cdot u_k + \frac{1}{2} \sum_{v=2}^{n} L_c \cdot y_{k,v} \cdot x_{k,v}) \end{aligned} \quad (2.35)$$

Hence,

$$\begin{aligned} \gamma_k(s',s) &= P(y_k|u_k) \cdot P(u_k) \\ &= B_k \cdot exp(\frac{1}{2} L_c \cdot y_{k,1} \cdot u_k + \frac{1}{2} \sum_{v=2}^{n} L_c \cdot y_{k,v} \cdot x_{k,v}) \\ &\quad \cdot A_k \cdot exp(\frac{1}{2} u_k \cdot L(u_k)) \\ &= A_k \cdot B_k \cdot exp(\frac{1}{2} L_c \cdot y_{k,1} \cdot u_k + \frac{1}{2} \sum_{v=2}^{n} L_c \cdot y_{k,v} \cdot x_{k,v} \\ &\quad + \frac{1}{2} u_k \cdot L(u_k)) \end{aligned} \quad (2.36)$$

The terms A_k and B_k in Equation (2.36) are equal for all transitions from level $k-1$ to level k and hence will cancel out in the ratio of Equation (2.31). Therefore, the branch transition operation to be used in Equation (2.29) and Equation (2.30), is reduced to the expression

$$exp(\frac{1}{2}u_k(L_c \cdot y_{k,1} + L(u_k))) \cdot \gamma_k^{(e)}(s', s) \tag{2.37}$$

with

$$\gamma_k^{(e)}(s', s) = exp(\frac{1}{2}\sum_{v=2}^{n} L_c \cdot y_{k,v} \cdot x_{k,v}) \tag{2.38}$$

Thus, the log-likelihood ratio becomes

$$L(\hat{u}_k) = L_c \cdot y_{k,1} + L(u_k) + \ln \frac{\sum_{u_k=+1}^{(s',s)} \gamma_k^{(e)}(s', s) \cdot \alpha_{k-1}(s') \cdot \beta_k(s)}{\sum_{u_k=-1}^{(s',s)} \gamma_k^{(e)}(s', s) \cdot \alpha_{k-1}(s') \cdot \beta_k(s)} \tag{2.39}$$

As discussed following the Equation (2.23), for any random bit in the information sequence:

$$L(\hat{u}_k) = L_c \cdot y_{k,1} + L(u_k) + L_e(\hat{u}_k) \tag{2.40}$$

the extrinsic information can be calculated as:

$$L_e(\hat{u}_k) = L(\hat{u}_k) - L_c \cdot y_{k,1} - L(u_k) \tag{2.41}$$

or

$$L_e(\hat{u}_k) = \ln \frac{\sum_{u_k=+1}^{(s',s)} \gamma_k^{(e)}(s', s) \cdot \alpha_{k-1}(s') \cdot \beta_k(s)}{\sum_{u_k=-1}^{(s',s)} \gamma_k^{(e)}(s', s) \cdot \alpha_{k-1}(s') \cdot \beta_k(s)} \tag{2.42}$$

2.2.3.2 Log-MAP Algorithm. The Log-MAP algorithm is a transformation of MAP, which has equivalent performance without its problems in practical implementation. It works in the logarithmic domain, where multiplication is converted to addition. The following are the calculations of branch transition probabilities and the forward/backward recursion formulas:

$$\begin{aligned} \gamma_k^{LM}(s', s) &= \ln \gamma_k(s', s) \\ &= \frac{1}{2}L_c \cdot y_{k,1} \cdot u_k + \frac{1}{2}\sum_{v=2}^{n} L_c \cdot y_{k,v} \cdot x_{k,v} + \frac{1}{2}u_k \cdot L(u_k) \end{aligned} \tag{2.43}$$

$$\begin{aligned} \alpha_k^{LM}(s) &= \ln \alpha_k(s) \\ &= \ln(\sum_{s'} e^{\gamma_k^{LM}(s',s)} \cdot e^{\alpha_{k-1}^{LM}(s)}) \\ &= \ln(\sum_{s'} e^{\gamma_k^{LM}(s',s) + \alpha_{k-1}^{LM}(s)}) \end{aligned} \tag{2.44}$$

$$\begin{aligned}
\beta_{k-1}^{LM}(s') &= \ln \beta_{k-1}(s') \\
&= \ln(\sum_{s} e^{\gamma_k^{LM}(s',s)} \cdot e^{\beta_k^{LM}(s')}) \\
&= \ln(\sum_{s} e^{\gamma_k^{LM}(s',s)+\beta_k^{LM}(s')})
\end{aligned} \tag{2.45}$$

Therefore, the log-likelihood ratio is given by

$$\begin{aligned}
L(\hat{u}_k) &= \ln \frac{\sum_{u_k=+1}^{(s',s)} e^{\gamma_k^{LM}(s',s)} \cdot e^{\alpha_{k-1}^{LM}(s)} \cdot e^{\beta_k^{LM}(s')}}{\sum u_{k=-1}^{(s',s)} e^{\gamma_k^{LM}(s',s)} \cdot e^{\alpha_{k-1}^{LM}(s)} \cdot e^{\beta_k^{LM}(s')}} \\
&= \ln \frac{\sum_{u_k=+1}^{(s',s)} e^{\gamma_k^{LM}(s',s)+\alpha_{k-1}^{LM}(s)+\beta_k^{LM}(s')}}{\sum_{u_k=-1}^{(s',s)} e^{\gamma_k^{LM}(s',s)+\alpha_{k-1}^{LM}(s)+\beta_k^{LM}(s')}} \\
&= \ln(\sum_{u_k=+1}^{(s',s)} e^{\gamma_k^{LM}(s',s)+\alpha_{k-1}^{LM}(s)+\beta_k^{LM}(s')}) \\
&\quad - \ln(\sum_{u_k=-1}^{(s',s)} e^{\gamma_k^{LM}(s',s)+\alpha_{k-1}^{LM}(s)+\beta_k^{LM}(s')})
\end{aligned} \tag{2.46}$$

2.2.3.3 Max-function. Define

$$E(x, y) = ln(e^x + e^y) \tag{2.47}$$

$$\begin{aligned}
\ln(e^x + e^y) &= \ln e^x + \ln(e^x + e^y) - \ln e^x \\
&= x + \ln \frac{e^x + e^y}{e^x} \\
&= x + \ln(1 + e^{y-x})
\end{aligned} \tag{2.48}$$

Similar way

$$\begin{aligned}
\ln(e^x + e^y) &= \ln e^y + \ln(e^x + e^y) - \ln e^y \\
&= y + \ln \frac{e^x + e^y}{e^y} \\
&= y + \ln(1 + e^{x-y})
\end{aligned} \tag{2.49}$$

Hence

$$\begin{aligned}
E(x, y) &= \ln(e^x + e^y) \\
&= \max(x, y) + \ln(1 + e^{-|x-y|})
\end{aligned} \tag{2.50}$$

and take

$$E(x, y) = \ln(e^x + e^y) \approx \max(x, y) \tag{2.51}$$

Similarly,

$$E(x, y, z, w) = \ln(e^x + e^y + e^z + e^w) \tag{2.52}$$

$$\begin{aligned} \ln(e^x + e^y + e^z + e^w) &= \ln e^x + \ln(e^x + e^y + e^z + e^w) - \ln e^x \\ &= x + \ln \frac{e^x + e^y + e^z + e^w}{e^x} \\ &= x + \ln(1 + e^{y-x} + e^{z-x} + e^{w-x}) \end{aligned} \tag{2.53}$$

or

$$\ln(e^x + e^y + e^z + e^w) = y + \ln(1 + e^{x-y} + e^{z-y} + e^{w-y}) \tag{2.54}$$

or

$$\ln(e^x + e^y + e^z + e^w) = z + \ln(1 + e^{x-z} + e^{y-z} + e^{w-z}) \tag{2.55}$$

or

$$\ln(e^x + e^y + e^z + e^w) = w + \ln(1 + e^{x-w} + e^{y-w} + e^{z-w}) \tag{2.56}$$

Hence

$$\begin{aligned} E(x, y, z, w) &= \max(x, y, z, w) \\ &\quad + \ln(e^{x-\max(x,y,z,w)} + e^{y-\max(x,y,z,w)} \\ &\quad + e^{z-\max(x,y,z,w)} + e^{w-\max(x,y,z,w)}) \\ &\approx \max(x, y, z, w) \end{aligned} \tag{2.57}$$

In general, we have

$$\begin{aligned} E(x_1, x_2, \ldots, x_k) &= \ln \sum_{i=1}^{k} e^{x_i} \\ &= \ln(e^{x_1} + e^{x_2} + \cdots + e^{x_{k-2}} + e^{x_{k-1}} + e^{x_k}) \\ &= \ln(e^{x_1} + e^{x_2} + \cdots + e^{x_{k-2}} + e^{\ln(e^{x_{k-1}} + e^{x_k})}) \\ &= \ln(e^{x_1} + e^{x_2} + \cdots + e^{x_{k-2}} + e^{E(x_{k-1}, x_k)}) \\ &= \ln(e^{x_1} + e^{x_2} + \cdots + e^{\ln(e^{x_{k-2}} + e^{E(x_{k-1}, x_k)})}) \\ &= \ln(e^{x_1} + e^{x_2} + \cdots + e^{E(x_{k-2}, E(x_{k-1}, x_k))}) \\ &= E(x_1, E(x_2, \cdots, E(x_{k-2}, E(x_{k-1}, x_k)))) \end{aligned} \tag{2.58}$$

Hence

$$E(x_1, x_2, \ldots, x_k) = \ln \sum_{i=1}^{k} e^{x_i} = \max(x_i) + \ln \sum_{i=1}^{k} e^{x_i - \max(x_i)} \tag{2.59}$$

2.2.3.4 Max-Log-MAP Algorithm. With max-function, the Log-MAP algorithm becomes Max-Log-MAP algorithm resulting in some degradation in the performance, but, with a drastic reduction in computational complexity. The correction term that compensates the degradation in the performance will be discussed in Chapter 8.

$$\alpha_k^{MLM}(s) = \max([\gamma_{u_k=+1}^{LM}(s', s) + \alpha_{k-1}^{LM}(s')], [\gamma_{u_k=-1}^{LM}(s', s) + \alpha_{k-1}^{LM}(s')]) \tag{2.60}$$

$$\beta_{k-1}^{MLM}(s') = \max([\gamma_{u_k=+1}^{LM}(s', s) + \beta_k^{LM}(s)], [\gamma_{u_k=-1}^{LM}(s', s) + \beta_k^{LM}(s)]) \tag{2.61}$$

$$\begin{aligned} L(\hat{u}_k) = & \max_{u_k=+1}^{(s',s)}([\gamma_k^{LM}(s', s) + \alpha_{k-1}^{MLM}(s) + \beta_k^{MLM}(s')], \\ & [\gamma_k^{LM}(s', s) + \alpha_k^{MLM}(s) + \beta_k^{MLM}(s')]) \\ & - \max_{u_k=-1}^{(s',s)}([\gamma_k^{LM}(s', s) + \alpha_{k-1}^{MLM}(s) + \beta_k^{MLM}(s')], \\ & [\gamma_k^{LM}(s', s) + \alpha_k^{MLM}(s) + \beta_k^{MLM}(s')]) \end{aligned} \tag{2.62}$$

2.2.3.5 SOVA Algorithm. *Soft-Output Viterbi Algorithm* (SOVA) accepts and delivers soft sample values. It delivers not only the most likely path sequence in a finite-state Markov chain, but also either the a posteriori probability for each bit or a reliability value [54]. The VA in its most general form is a maximum *a posteriori* probability sequence estimator [18], which finds the maximum *a posteriori* probability path over the trellis diagram given the received sequence $Y = \{y_k\}_{k=1}^N$, where the block length is N. A path in the trellis can be represented by a state sequence $S = \{s_1, s_2, ..., s_N\}$, which indicates the trellis path starting at state s_1, passing through every state s_k at time k, and terminating at state s_N. The VA finds the trellis path or state sequence S so that the *a posteriori* probability $P(S|Y)$ is maximized.

Assume that the state sequence S is a Markov sequence. Since the received sequence $Y = \{y_k\}_{k=1}^N$ does not depend on the selection of the trellis path S, and $P(S|Y) = p(S,Y)/p(Y)$, at time k, we can equivalently maximize

$$p(S_k, Y_k) = p(S_{k-1}, Y_{k-1})P(u_k)p(y_k|s', s) \tag{2.63}$$

where $S_k = \{s_1, s_2, ..., s_k\}$, $Y_k = \{y_1, y_2, ..., y_k\}$, $s' = s_{k-1}$, $s = s_k$, and u_k is the source bit or symbol corresponding to the state transition $s' \rightarrow s$ of trellis path S_k.

The path metric $M_k(S_k)$ associated with the trellis path S_k is defined as

$$M_k(S_k) = \ln(p(S_k, Y_k)). \tag{2.64}$$

Obviously,

$$p(S_k, Y_k) = \exp(M_k(S_k)). \tag{2.65}$$

Substituting Equation (2.63) into Equation (2.64) gives

$$M_k(S_k) = M_{k-1}(S_{k-1}) + \ln P(u_k) + \ln(p(y_k|s', s)). \tag{2.66}$$

Using the notation $\lambda = \log(P)$ [55], Equation (2.66) becomes

$$M_k(S_k) = M_{k-1}(S_{k-1}) + \lambda(u_k) + \lambda(y_k|s', s). \tag{2.67}$$

where $\lambda(u_k)$ is the *a priori* information of the source symbol u_k, and $\lambda(y_k|s', s)$ is the branch metric corresponding to the state transition $s' \rightarrow s$ given the received signal y_k. At time k, for each state s, the path metrics for all possible paths terminating at state s are calculated. Only the maximal path metric is saved and the corresponding path is the survivor path.

For binary RSC code with rate $1/n$ described in Section 2.2.3.1, the metric calculation can be simplified as follows [7]:

$$M_k(S_k) = M_{k-1}(S_{k-1}) + \frac{1}{2}L(u_k)u_k + \frac{1}{2}\sum_{v=1}^{n} L_c y_{k,v} x_{k,v} \tag{2.68}$$

furthermore, for the systematic codes we have

$$M_k(S_k) = M_{k-1}(S_{k-1}) + \frac{1}{2}L(u_k)u_k + \frac{1}{2}L_c y_{k,1} u_k + \frac{1}{2}\sum_{v=2}^{n} L_c y_{k,v} x_{k,v} \tag{2.69}$$

The sum is over the indices v with nonpunctured coded bits. The following figure illustrates the procedure of updating the soft information.

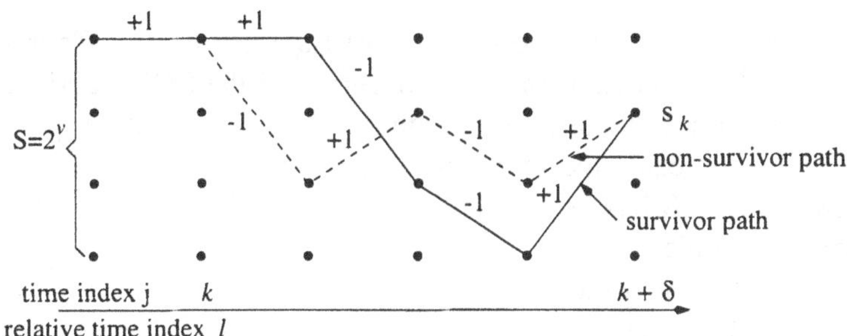

Figure 2.5. Update of the Soft Information for the Coded Bits. Example of the SOVA.

The SOVA can be implemented in the trace back mode using the log-likelihood algebra, $L(u_k) = \ln \frac{P(u_k=+1)}{P(u_k=-1)}$, to represent the soft output in the same way as the binary MAP algorithm. We wish to obtain the soft output for bit $\hat{u}_k$, which the VA decides after a delay δ. The VA proceeds in the usual way by calculating the metrics for the path. For each state it selects the path with the larger metric $M_k(S_k)$. The soft output of the SOVA is approximated as [7]:

$$L(\hat{u}_k) \approx \hat{u}_k \cdot \min_{l=k,\ldots,\delta} \Delta_k^l \tag{2.70}$$

where $\Delta_k^l = M_l(S_l^{(survivor)(\hat{u}_k)}) - M_l(S_l^{(non-survivor)(-\hat{u}_k)})$ is the metric difference at time l, and $S_l^{(survivor)(\hat{u}_k)}$ is the survivor trellis path terminated at s_l which contains the branch $s' \rightarrow s$ with its corresponding $u_k = \hat{u}_k$; and $S_l^{(non-survivor)(-\hat{u}_k)}$ is the non-survivor path terminated at s_l which contains the branch $s' \rightarrow s$ with its corresponding $u_k = -\hat{u}_k$. Then the probability $P(correct)$ that the path decision of the survivor was correct at time l is

$$\begin{aligned} P(correct) &= \frac{P(survivor\ path)}{P(survivor\ path) + P(non - survivor\ path)} \\ &= \frac{\exp(M_l(S_l^{(survivor)(\hat{u}_k)}))}{\exp(M_l(S_l^{(survivor)(\hat{u}_k)})) + \exp(M_l(S_l^{(non-survivor)(-\hat{u}_k)}))} \\ &= \frac{\exp(\Delta_k^l)}{1 + \exp(\Delta_k^l)} \end{aligned}$$

Therefore, the likelihood ratio or "soft value" of this binary path decision is Δ_k^l, because

$$\ln \frac{P(correct)}{1 - P(correct)} = \Delta_k^l$$

Furthermore, the SOVA output in its approximate version in Equation (2.70) has the format [7]

$$L_{SOVA}(\hat{u}_k) = L_c y_{k,1} + L(u_k) + L_e(\hat{u}_k) \tag{2.71}$$

and preserves the desired additive structure of Equation (2.23). Consequently, we subtract the input values from the soft output of the SOVA and obtain the extrinsic information to be used in the metrics of the succeeding decoder (see Figure 2.1). In this case, the extrinsic term in 2.71 is weakly correlated to the other two terms. It has been shown that for small memories the SOVA is roughly half as complex as the Log-MAP algorithm [56].

Turbo decoder entails much higher decoding complexity than conventional channel decoders. To ensure its true success, high-speed decoder implementation is much needed for turbo code. Therefor, the suboptimal algorithm Max-Log-MAP and SOVA are accepted in practice.

2.3. Parallel Concatenation

The original turbo code [6] is the combination of two parallel *Recursive Systematic Convolutional* (RSC) codes concatenated by a pseudo-random interleaver, and an iterative MAP decoder. The turbo coding/decoding principle is illustrated in Figure 2.6. Π represents the interleaver between Encoder 1 and Encoder 2 and Π^{-1} represents deinterleaver between Decoder 2 and Decoder 1.

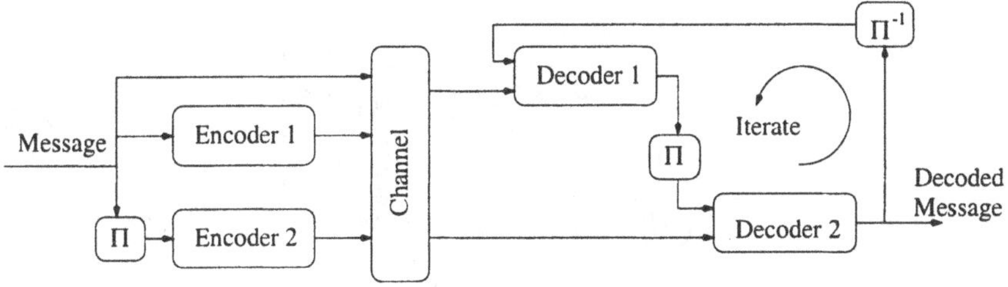

Figure 2.6. The Turbo Coding/Decoding Principle

According to the turbo coding principle, the turbo code design issues include component code design, trellis termination method, interleaving strategy and implementation complexity based on the system design space for turbo codes shown in Figure 2.7 [61].

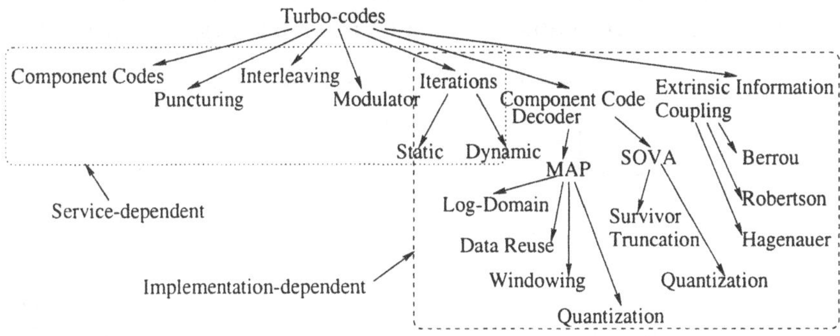

Figure 2.7. System Design Space

The system design space comprises a service-dependent and an implementation-dependent part [61]. The components of the turbo encoder directly define the service-dependent part of the system design space: component codes, the interleaver, the puncturer and the modulator. Though the required number of iterations is implementation-dependent, this number may also depend on the service to realize different qualities of service. For static iterations, the number of iterations is predetermined. Dynamic iterations depend on when the soft-output values $L^2(\hat{u})$ stabilize.

2.3.1 The Component Encoder with Binary Codes

A general binary convolutional turbo encoder structure using two component encoders is illustrated in Figure 2.8 as an example. It consists of three basic building blocks: an interleaver Π, the component encoders, and a puncturing device with a multiplexing unit to compose the codeword. The interleaver is a device that re-orders the symbols in its input sequence.

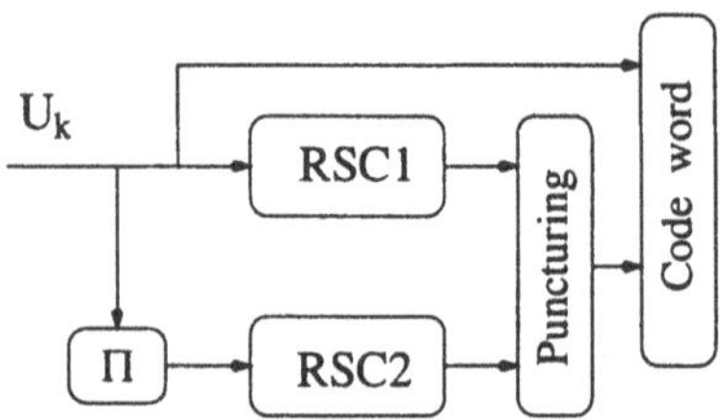

Figure 2.8. Encoder Block Diagram (Binary)

The Component encoders are RSC encoders, *i.e.,* systematic convolutional encoders with feedback. Such an encoder with two memory elements is depicted in Figure 2.9. For systematic codes, the information sequence is part of the codeword, which corresponds to a direct connection from the input to one of the outputs. For each input bit, the encoder generates two codeword bits: the systematic bit and the parity bit. Thus, the code rate is 1/2 and the encoder input and output bits are denoted U_k and $(X_{k,1} = U_k,\ X_{k,2})$, respectively.

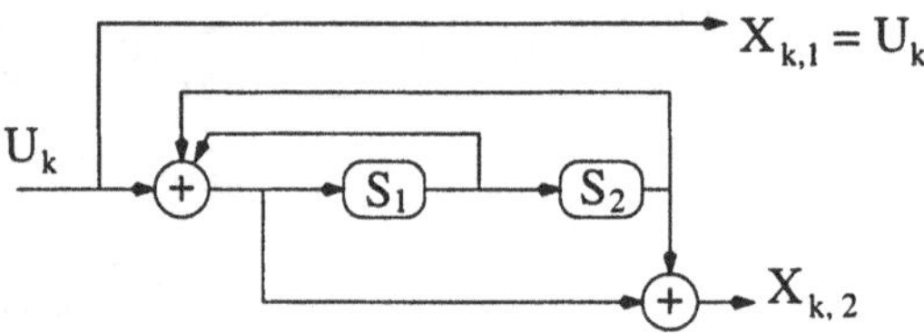

Figure 2.9. Recursive systematic convolutional encoder with feedback for rate 1/2 code with memory 2. The generator polynomials are $g_0(D) = 1 + D + D^2$ and $g_1(D) = 1 + D^2$.

If the generator matrix of a non-recursive convolutional encoder with rate 1/n is given by

$$G(D) = (g_0(D),\ g_1(D),\ ...,\ g_{n-1}(D)) \tag{2.72}$$

the recursive encoder will be defined by,

$$G_{sys}(D) = (1,\ \frac{g_1(D)}{g_0(D)},\ ...,\ \frac{g_{n-1}(D)}{g_0(D)}) \tag{2.73}$$

Since the performance of any binary code is dominated by its free distance d_f (the minimum Hamming distance between codewords, which coincides with the minimum Hamming weight of a nonzero codeword for linear codes) and its multiplicity [62], the optimal-recursive component encoders should have maximum effective free distance and minimum multiplicity to achieve a good performance. Furthermore, to achieve a good performance, it is also important that the component codes be recursive.

In the design of convolutional codes, one advantage of systematic codes is that encoding is somewhat simpler than for the non-systematic codes and less hardware is required. Another advantage is that no inverting circuitry is needed for recovering the information sequence from the codeword [34].

2.3.2 Interleaving

Interleaving is the process of rearranging the ordering of an information sequence in a one-to-one deterministic way before the application of the second component code in a turbo coding scheme. The inverse of this process is called deinterleaving which restores the received sequence to its original order. Interleaving is a practical technique to enhance the error correcting capability of the coding schemes [60]. It plays an important role in achieving good performance in turbo coding schemes.

Constructing a long block code from short memory convolutional codes using the interleaver results in the creation of codes with good distance properties, which can be efficiently decoded through iterative decoding [63]. The interleaver breaks low weight input sequences, and hence increases the code's free Hamming distance or reduces the number of codewords with small distance in the code distance spectrum. On the other hand, the interleaver spreads out burst errors through providing "scrambled" information data to the second component encoder, and at the decoder, decorrelates the inputs to the two component decoders so that an iterative sub-optimum decoding algorithm based on "uncorrelated" information exchange between the two component decoders can be applied. For example, after correction of some of the errors in the first component decoder, some of the remaining errors can be spread by the interleaver such that they become correctable in the other decoder. By increasing the number of iterations in the decoding process, the bit error probability approaches that of the maximum likelihood decoder. Typically, the performance of a turbo code is improved when the interleaver size is increased, which has a positive influence on both the code properties and iterative decoding performance.

A key component of turbo code is the interleaver whose design is essential for achieving high performance and is of interest to many turbo code researchers. Many interleaving strategies have been proposed, including block interleavers, Odd-Even block interleavers, block helical simile interleavers; Convolutional interleavers and Cyclic shift interleavers; Random interleavers including pseudo-random interleaver, Uniform and Non-uniform interleavers, S-random interleavers; Code matched interleavers, Relative prime interleavers; Golden interleavers, *etc.* [66], [64], [65], [67], [68], [38], [69], [70], [71], [72], [73], [74], [75] and [76].

2.3.3 Trellis Termination

As mentioned above, the performance of a code is highly dependent on its Hamming distance spectrum. For convolutional turbo codes, the Hamming distances between the codewords are the result of taking different paths through the trellis. In principle, the larger the number of trellis transitions in which the two paths differ, the larger is the possible Hamming distance between the corresponding codewords. It is thus desirable that the shortest possible detour from a trellis path is as long as possible, to ensure a large Hamming distance between the two codewords that correspond to the two paths. However, in practice, convolutional turbo codes are truncated at some point in order to encode the information sequence block-by-block. If no precautions are taken before the truncation, each of the encoder states is a valid ending state and thus the shortest possible difference between the two trellis paths is made up of only one trellis transition. Naturally, this procedure may result in very poor distance properties, with accompanying poor error correcting performance. This problem has been discussed in [77], [78], [79], [63], [80], [81] and [82].

Since the component codes are recursive, it is not possible to terminate the trellis by transmitting ν zero tail bits. The tail bits are not always zero, and depend on the state of the component encoder after encoding N information bits. Trellis termination forces the encoder to the all-zero state at the end of each block to make sure that the initial state for the next block is the all-zero state. This way, the shortest possible trellis detour does not change with truncation, and the distance spectrum is preserved.

Another approach to the problem of trellis truncation is tail-biting. With tail-biting, the encoder is initialized to the same state that it will end up in, after encoding the whole block. For feed-forward encoders tail-biting is readily obtained by inspection of the last ν bits in the input sequence, since these dictate the encoder ending state. The advantage of using tail-biting compared to trellis termination is that tail-biting does not require transmission of tail bits (the use of tail bits reduces the code rate and increases the transmission bandwidth). For large blocks, the rate-reduction imposed by tail-bits is small, often negligible. For small blocks, however, it may be significant. References [83], [84], [85], [86] and [87] address tail-biting.

2.3.4 Puncturing

The total rate of the two parallel concatenation codes without puncturing will be

$$\frac{1}{R} = \frac{1}{R_1} + \frac{1}{R_2} - 1 \tag{2.74}$$

Example 1: (basic rate 1/3) If two convolutional codes both have rate $R_1 = R_2 = 1/2$, then $R = 1/3$.

Example 2: (rate 1/6) If two different convolutional codes have rates $R_1 = 1/3$ and $R_2 = 1/4$, then $R = 1/6$.

Example 3: (higher rates) Higher rate turbo codes can be constructed from higher-rate convolutional codes. If two convolutional codes both have rate

$R_1 = R_2 = 3/4$, then $R = 3/5$. If two convolutional codes both have rate $R_1 = R_2 = 4/5$, then $R = 2/3$. It may be difficult to find a convolutional code with high R without a large number of states. Therefore, puncturing is a good solution to increase the code rate without a large number of states and the complexity of the codes is low.

Puncturing is the process of removing certain symbols/positions from the codeword, thereby reducing the codeword length and increasing the overall code rate. In the original turbo code proposal, Berrou *et al.* punctured half of the bits from each constituent encoder. Puncturing half of the systematic bits from each constituent encoder corresponds to sending all the systematic bits once, if the puncturing is properly performed. The overall code rate is $R = 1/2$. Furthermore, puncturing may have different effect for different choices of interleavers, and for different constituent encoders.

When puncturing is considered, for example, some output bits of v_0, v_1 and v_2 are deleted according to a chosen pattern defined by a puncturing matrix P. For instance, a rate 1/2 turbo code can be obtained by puncturing a rate 1/3 turbo code. Commonly used puncturing matrix is given by

$$\mathrm{P} = \begin{bmatrix} 1 & 1 \\ 1 & 0 \\ 0 & 1 \end{bmatrix}$$

where the puncturing period is 2. According to the puncturing matrix, the parity check digits from the two component encoders are alternately deleted. The punctured turbo code symbol at a given time consists of an information digit followed by a parity check digit which is alternately obtained from the first and the second component encoders.

2.3.5 Multiple Parallel Concatenation of Turbo Codes

In general, a parallel concatenated turbo codes can be constructed through multiple parallel concatenation. Figure 2.10 gives the principle of the encoding. Multilevel coding uses partition chains to encode different bits of the input stream with different codes, and is typically applicable only to systems with very large rate R.

2.4. Applications of Parallel Concatenated Turbo Codes

Turbo code, due to its excellent error correcting capability, has received much attention world wide and has been adopted by 3rd generation (3G) mobile communication standards such as 3GPP (*3rd Generation Partnership Project*), UTMS (*Universal mobile Telecommunication Service*), and CDMA2000 (*Code Division Multiple Access 2000*).

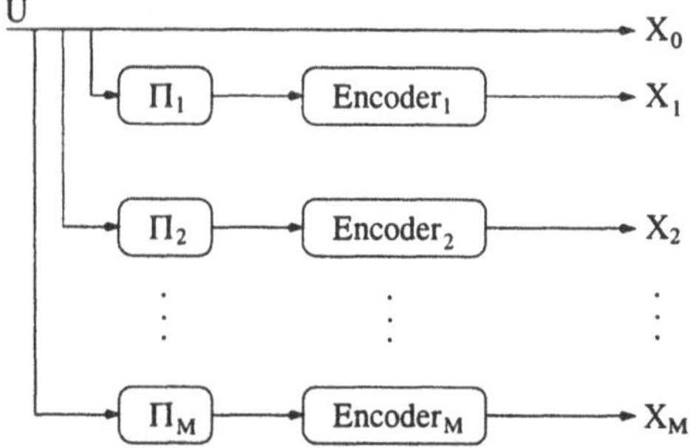

Figure 2.10. Multiple Parallel Concatenation Codes

2.4.1 Turbo Codes in 3GPP

The turbo coding scheme in 3GPP standard [37] is a PCCC with two 8-state constituent encoders and an internal interleaver. The coding rate of the turbo encoder is 1/3. The structure of the turbo encoder is illustrated in Figure 2.11. The transfer function of the 8-state constituent code for PCCC is:

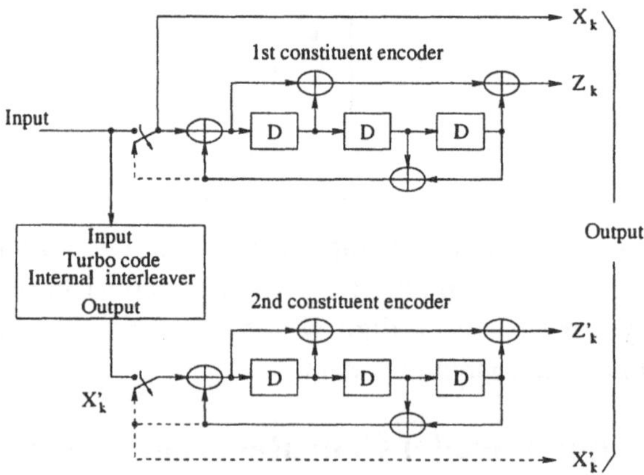

Figure 2.11. Structure of Rate 1/3 3GPP Turbo Encoder (dotted lines apply for trellis termination only)

$$G(D) = [1, \frac{g_1(D)}{g_0(D)}],$$

$$g_0(D) = 1 + D^2 + D^3,$$

$$g_1(D) = 1 + D + D^3.$$

The initial value of the shift registers of the 8-state constituent encoders shall be all zeros when starting to encode the input bits.

Output from the turbo encoder is

$$x_1,\ z_1,\ z_1',\ x_2,\ z_2,\ z_2',\ ...,\ x_K,\ z_K,\ z_K',$$

where x_1, x_2, ..., x_K are the bits input to the turbo encoder; K is the number of bits; z_1, z_2, ..., z_K and z_1', z_2', ..., z_k' are the bits output from first and second 8-state constituent encoders, respectively. The bits output from the internal interleaver are denoted by x_1', x_2', ..., x_K', and these bits are to be input to the second 8-state constituent encoder.

Figure 2.12 shows the BER performance of 3GPP turbo decoder implementated on a Texas Instruments DSP.

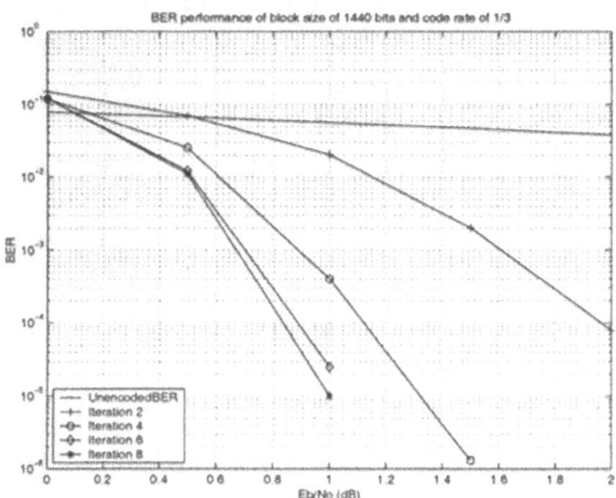

Figure 2.12. BER Performance of DSP Turbo Decoder using the Max-Log-MAP Algorithm

2.4.1.1 Trellis Termination for Turbo Encoder.

Trellis termination is performed by taking the tail bits from the shift register feedback after all information bits are encoded. Tail bits are padded after the encoding of information bits.

The first three tail bits shall be used to terminate the first constituent encoder (upper switch of Figure 2.11 in lower position) while the second constituent encoder is disabled. The last three tail bits shall be used to terminate the second constituent encoder (lower switch of Figure 2.11 in lower position) while the first constituent encoder is disabled.

The transmitted bits for trellis termination shall then bex_{K+1}, z_{K+1}, x_{K+2}, z_{K+2}, x_{K+3}, z_{K+3}, x'_{K+1}, z'_{K+1}, x'_{K+2}, z'_{K+2}, x'_{K+3}, z'_{K+3}

2.4.1.2 Turbo Code Internal Interleaver.

The operation of 3GPP turbo code internal interleaver is described in [37]. The bits are input to a rectangular matrix row-by-row and then padding, inter-row and intra-row permutations are performed and finally, they are read out column-by-column. The bit sequence x_1, x_2, ..., x_K input to the turbo code internal interleaver is written into the rectangular matrix whose size depends on the frame size. Using some lookup tables and equations, the intra-row and inter-row permutations reorder the information bit sequence to achieve better distance spectrum. Then, the output of the turbo code internal interleaver is the bit sequence read out column by column from the intra-row and inter-row permuted rectangular matrix with pruning.

2.4.2 Turbo Codes in CDMA2000

In CDMA2000 proposal, turbo codes are recommended for both forward and reverse supplemental channels in the 3rd generation of the *Wideband Code Division Multiple Access* (WCDMA) cellular mobile systems [The CDMA2000 ITU-R RTT Candidate Submission (0.18), July. 1998].

A common constituent code shall be used for the turbo codes of rate 1/2 and 1/4. The transfer function for the constituent coded shall be

$$G(D) = [1, \frac{n_0(D)}{d(D)}, \frac{n_1(D)}{d(D)}]$$

where $d(D) = 1 + D^2 + D^3$, $n_0(D) = 1 + D + D^3$, and $n_1(D) = 1 + D + D^2 + D^3$. The encoder structure is shown in Figure 2.13 and includes the termination.

For the third generation CDMA data services under development in the U.S., Europe, and Asia, the turbo interleaver design must be able to support many different block sizes from approximately 300 information bits up to 8192 information bits or more according to variable input data rate requirements [38]. The prunable interleavers are designed for each required block size. These are optimized block interleavers with pseudo-random readout that provide high performance even under severe pruning. The procedure of interleaving is described in [39] and Figure 2.14 shows the performance of the turbo code in CDMA2000 standard.

2.4.3 Turbo Codes for Deep Space Communications

The CCSDS (*Consultative Committee for Space Data Systems*) standard for deep-space telemetry has often represented a benchmark for new coding technologies. Recently, the old channel coding standard has been updated to include turbo codes [62]. The BER/FER performance of the CCSDS turbo code for low-medium E_b/N_0 has been largely studied: it has been pointed out that an additional coding gain of 2.5dB at $BER = 10^{-5}$ can be achieved by the rate 1/6 turbo code with respect to the old standard.

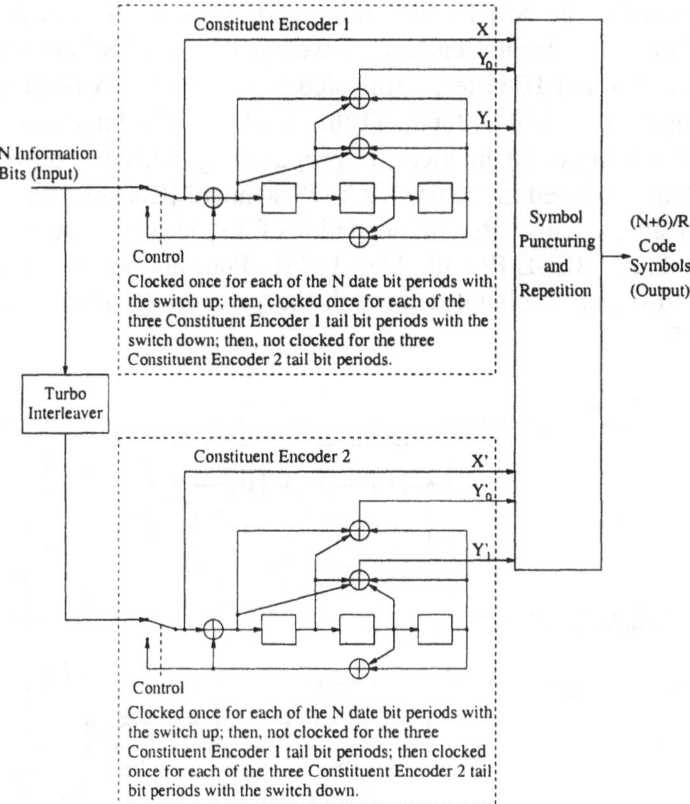

Figure 2.13. Turbo Encoder (CDMA2000)

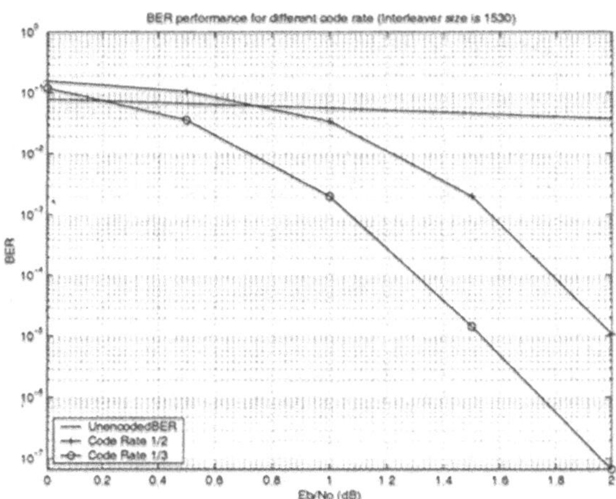

Figure 2.14. The Performance of CDMA2000

The encoder of the new CCSDS turbo code is depicted in Figure 2.15. It consists of the parallel concatenation of two equal binary RSC encoders E_1 and E_2 with rate 1/4 and 16 states, terminated in four steps. A block interleaver Π with length N = 1784, 3568, 7136, *or* 8920. The interleaver for all of these sizes are analytical interleavers proposed by Berrou and generated by the algorithm described in the new CCSDS standard [Consultative Committee for Space Data Systems, "Recommendations for space data systems, telemetry channel coding," BLUE BOOK, May 1998]. Four normal code rates 1/r, for r=2,3,4, and 6 can be obtained through the puncturing options described in Figure 2.15.

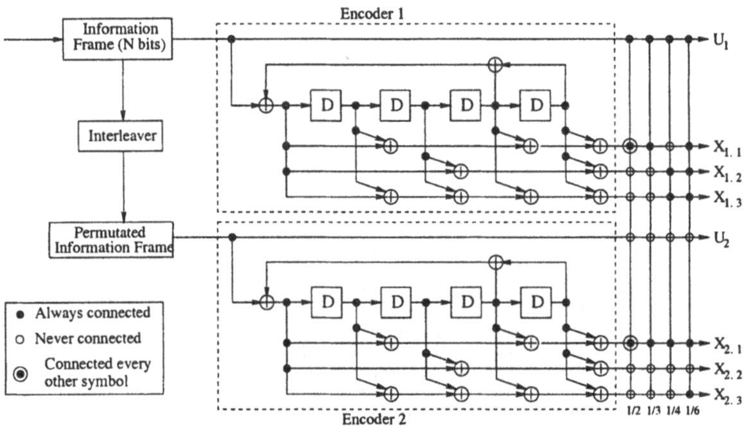

Figure 2.15. Encoder for the CCSDS Turbo Code

2.5. Serial Concatenation

A serial concatenated code was first conceived by Forney [46], see Figure 2.16. It was shown that the probability of error for serial concatenated codes decreases exponentially as the frame size increases at rates less than capacity while decoding complexity increases only algebraically.

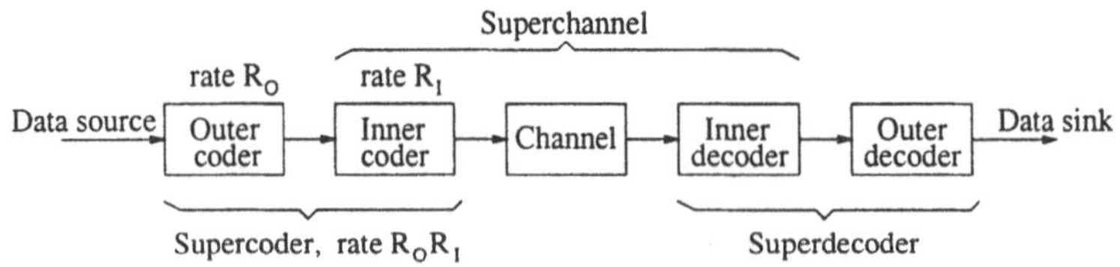

Figure 2.16. Serial Concatenation Codes

The outer encoder produces outer code symbols (or sequences of symbols) that enter the inner encoder as information symbols (sequences). The inner encoder, in turn, produces inner code symbols that are transmitted using some memoryless modulation. Forney showed that the optimal concatenated receiver consists of an inner detector producing a posteriori probabilities (APP) of the inner information symbols, which are then used in the outer detector.

The best known example is a *Reed-Solomon* (RS) outer code concatenated with a convolutional inner code separated by a symbol interleaver. A SCCC is the result of combining the features of serial concatenated codes with those of turbo codes. Unlike the symbol interleaver between RS and the convolutional code, a bit interleaver is used in SCCCs to introduce randomness.

2.5.1 Structure of SCCC

Using the same components as the turbo codes, such as constituent encoders, the interleaver, the puncturer, and the soft-input soft-output MAP decoders, another type of concatenated codes, serial concatenated convolutional codes (SCCCs), were proposed [59]. The good performance of SCCCs has led to a lot of investigations and applications in the coding field.

The basic structure of a SCCC encoder is shown in Figure 2.17, where u and c represent the input output symbol respectively. The information bits are encoded by the outer encoder, whose output sequence is passed to the bit interleaver. The bit interleaver permutes the output of the outer encoder and passes it as the input to the inner encoder. The output of the inner encoder is transmitted through the channel. The general structure of a SCCC encoder encompasses two or more serially cascaded constituent encoders separated by one or more interleavers.

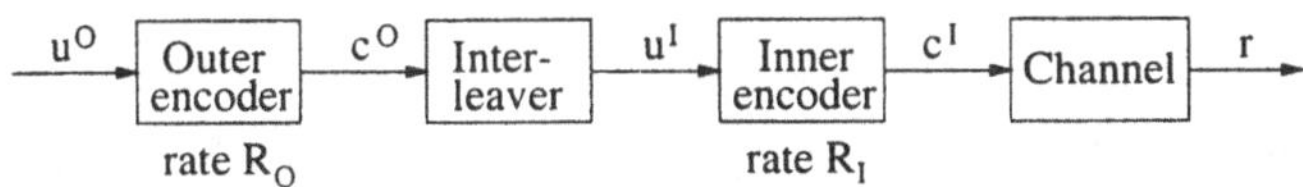

Figure 2.17. Encoder Structure of SCCC.

The outer encoder can be block code, convolutional code or recursive systematic convolutional code with any code rate. The inner code can be any code or TCM (*Trellis Coded Modulation*) or CPM (*Continuous-phase modulation*), or any other modulation scheme with memory. The *serial concatenated Trellis Coded Modulation* (SCTCM) with turbo codes is discussed in Chapter 4.

The code rate of the SCCC is the product of the outer code rate and the inner code rate, $R = R_O R_I$. Just as for PCCC, SCCC also creates an overall code trellis with a huge number of states because of the bit interleaver; however, it can be decoded with a relatively simple iterative MAP decoding procedure.

2.5.2 Decoding Procedure of Serial Concatenation Codes

Figure 2.18 shows a block diagram of the decoder for a serial concatenation code. To decode a SCCC with SISO, the following procedure will be employed. Here the superscripts I and O denote the inner and outer code, respectively. The algorithm is the MAP described in subsection 2.2.3.1, however, it can be extended to any other algorithm such as Log-MAP, Max-Log-MAP and SOVA.

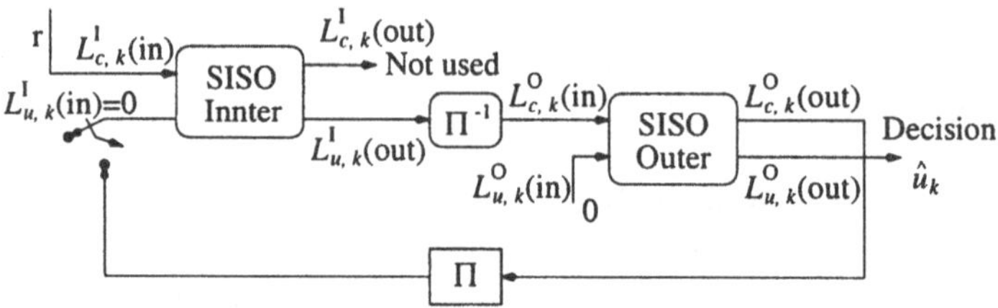

Figure 2.18. Serially Concatenated Convolutional Code with Iterative Decoding and General SISO Module.

The decoding procedure is as follows:

1 Scale the received signal using Equation (2.9) to get the LLR input $L^I_{c,k}(in)$ for the codeword bits of the inner code. For the first iteration, initialize $L^I_{u,k}(in) = 0$.

2 Inner decoder: Calculate path metrics $\alpha^I_k(s)$ and $\beta^I_k(s)$ with Equations (2.29) and (2.30). Then use Equations (2.31) and (2.41) to compute the extrinsic information $L^I_{u,k}(out)$ for the information bits of the inner code.

3 Deinterleave the stream of $L^I_{u,k}(out)$ to be the LLR input $L^O_{c,k}(in)$ for the codeword bits of the outer code.

4 Outer decoder: First calculate path metrics $\alpha^O_k(s)$ and $\beta^O_k(s)$ using Equations 2.29 and 2.30. Set $L^O_{u,k}(in) \equiv 0$, and it can be dropped from the computation. Then,

- if the number of iterations is less than the allowed number of iterations, use Equation (2.31) and 2.41 to calculate the extrinsic information $L^O_{c,k}(out)$ for the codeword bits of the outer code. Interleave the stream of $L^O_{c,k}(out)$ to be the LLR input $\lambda^I_{u,k}(out)$ of the inner code. Increment the iteration number, go back to Step (2) and start the next iteration.

- if the maximum number of iterations is reached, use Equation (2.31) (with $L^O_{u,k}(in) \equiv 0$) to obtain the complete information $L_{\hat{u},k}(out)$ of the outer code. Make decision on the transmitted bits as follows:

$$\hat{u}_k = \begin{cases} 1, & when \;\; L_{\hat{u},k}(out) > 0 \\ 0, & when \;\; L_{\hat{u},k}(out) < 0 \end{cases}$$

output the hard decisions for the current frame, and go to Step (1) to decode the next frame.

In comparison to PCCC whose $P_e(e)$ decrease at the rate of N^{-1}, where N is the interleaver size, $P_e(e)$ of SCCC decreases at a faster rate, e.g., N^{-2}, N^{-3}, ..., as N increases. The error floor associated with PCCC, where the bit error probability flattens, is eliminated by SCCC as a result. A disadvantage of the SCCC is that it is computationally more complex than the PCCC with constituent codes of the same memory size. Also, SCCC tends to have a higher bit error probability than PCCC at low SNR [40].

2.6. Summary

The fundamental principles behind binary convolutional turbo coding have been introduced, including the component codes, the interleaving, the trellis termination, the puncturing and the principle of iterative decoding. The central components of a turbo code encoder are the RSC encoders and the interleaver that links them in parallel by re-ordering the bits in the information sequence before they enter the second constituent encoder. Both optimal algorithms, MAP and Log-MAP and, suboptimal algorithms, Max-Log-MAP and SOVA depend on the tools, Log-likelihood Algebra, Soft channel output and the principle of iterative decoding algorithm. Turbo codes, due to their excellent error correcting capability, are being considered for the 3rd generation (3G) mobile communication standards, 3GPP, UTMS, and CDMA2000. Both parallel concatenation and serial concatenation were discussed and their decoding procedures were given.

The parallel concatenation is implemented by interleaving, *i.e.*, re-ordering the information sequence before it is input to the second component encoder. The two most critical parts of a turbo code encoder are, thus the interleaver and the component encoders. Other essential aspects of a turbo encoder are trellis termination and puncturing. Trellis termination is an issue when dealing with data packets where truncation is necessary at some point of the trellis. Puncturing is the process of excluding bits from the outputs of the component encoders, so the concatenated transmitted sequence is a decimated version of the encoder output.

Chapter 3

NON-BINARY TURBO CODES: DVB/RCS STANDARD

Double-binary elementary codes provide better error-correcting performance than binary codes for equivalent implementation complexity [89]. And also, a parallel concatenation of *Circular Recursive Systematic Convolutional* codes (CRSC) [90] makes convolutional turbo codes efficient for coding of data cells in blocks. The double-binary CRSC codes were adopted in the DVB-RCS standard for their excellent performance as an alternative to the conventional scheme consisting of the concatenation of a convolutional code and a RS code.

The codes investigated in this chapter are constructed via parallel concatenation of double-binary CRSC codes by a non-uniform interleaver. Circular coding is a kind of "tail-biting" technique that avoids reducing the code rate and increasing the transmission bandwidth. The influence of puncturing and suboptimal decoding algorithm, Max-Log-MAP algorithm, are less significant with double-binary turbo codes than with binary turbo codes. Using double-binary codes, the latency of the decoder is halved. Double-binary CRSC code could be easily adopted for many applications, for various block sizes and code rates, with retaining excellent coding gains.

3.1. Design of Double-binary CRSC Codes

For efficient convolutional turbo coding, the number of memory elements is a key consideration since the component codes with small constraint lengths ensure convergence at very low signal to noise ratios and the correlation effects are minimized [88]. Moreover, reasonable constraint lengths make hardware implementation on a single integrated circuit possible since the material complexity of the decoder grows exponentially with the code memory. The solution chosen uses component codes with memory $v = 3$. The encoder structure of the double-binary component codes is depicted in Figure 3.1.

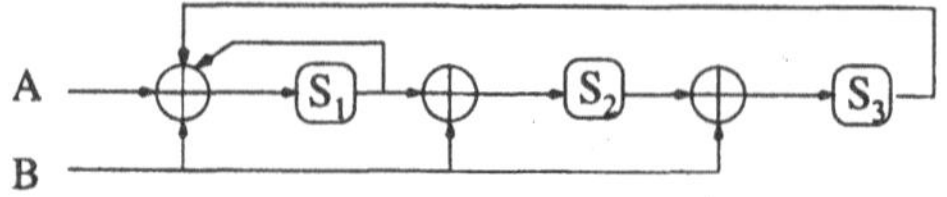

Figure 3.1. Recursive Convolutional (Double-binary) Encoder with Memory $v = 3$. The output, which is not relevant to the operation of the register, has been omitted

3.1.1 Two-level Permutation (Interleaving)

The performance of PCCC at low error rates, is essentially governed by the permutation that links the two component codes. The simplest way to achieve interleaving in a block is to adopt uniform or regular interleaving: data are written row-wise and read column-wise in a rectangular matrix. This kind of permutation behaves very well towards error patterns with weight 2 or 3, but is very sensitive to square or rectangular error patterns, as explained in [9]. Classically, in order to increase the distances given by rectangular error patterns, non-uniformity is introduced in the permutation relations. Many proposals have been made in this direction, especially for the UMTS application. The CCSDS turbo code standard may also be cited as an example of non-uniform permutation. However, the disorder that is introduced with non-uniformity can affect the scattering properties concerning weight 2 or 3 error patterns [36].

With double-binary codes, non-uniformity can be introduced without any repercussion on the good scattering properties of the regular interleaving [88]. The principle involves introducing local disorder into the data couples (two bits), for example (A, B) becoming (B, A) - or (B, A+B), etc. - periodically before the second encoding. This helps to avoid many error patterns. Therefore, this appears to be a significant gain in the search for large minimum distance.

3.1.2 Circular Recursive Systematic Convolutional (CRSC) Codes

For block-oriented encoding, convolutional turbo codes have to be truncated at some point. This will result in a degradation of performance if no precaution is taken. It is easy to know the initial state of the trellis as encoder is generally forced into the "all zero" state at the beginning of the encoding. However, the decoder has no special information available regarding the final state of the trellis. There are many approaches for solving this problem, for example, forcing the encoder state at the end of the encoding phase to a known state for one or all of the component encoders. Tail bits are used to "close" the trellises and are then sent to the decoder. This method presents two major drawbacks. First, minimum weight w_{min} is no longer equal to the original w_{min} for all information data[1], since, at the end of each block, the second "1" bringing the encoder back to the "all zero" state may be a part of the tail bits. In this case, turbo decoding is handicapped if tail bits are not encoded another time. The second problem is that the spectral efficiency of the transmission is degraded and the degradation is more for shorter blocks.

With circular convolutional codes, the encoder retrieves the initial state at the end of the encoding operation. Trellis can, therefore, be seen as a circle and decoding may be initialized everywhere on this circle. This technique, well known for non recursive codes (the so-called "tail-biting"), has been adapted to the specificity of the recursive codes [90]. Adopting circular coding avoids the degradation of spectral efficiency that occurs when the encoder is forced into a known state by the addition of tail bits [36].

3.1.3 Circular States (Tail-biting) Principle

Circular coding ensures that, at the end of the encoding operation, the encoder retrieves the initial state, so that data encoding may be represented by a circular trellis. The existence of such a state, called the circular state S_c, is ensured when the size of the encoded data block, N, is not a multiple of the period of the encoder's recursive generator. The value of the circulation state depends on the contents of the sequence to be encoded and determining S_c requires a pre-encoding operation: first, the encoder is initialized in the "all zero" state. The data sequence is encoded once, leading to a final state S_N^0. Next, we find S_c from the final state S_N^0 as explained below.

In practice, for example, in DVB standard [28], the relationship between S_c and S_N^0 is provided by a small combinatorial operator with $v = 3$ input and output bits. To perform a complete encoding operation of the data sequence, two circulation states have to be determined, one for each component encoder, and the sequence has to be encoded four times instead of twice.

Let us consider a recursive convolutional encoder, for instance, the encoder depicted in Figure 3.1 [90]. At time k, register state S_k is a function of the previous state S_{k-1} and the input vector X_k. Let G be the generator matrix of the considered code. State S_k and S_{k-1} are linked by the following recursion relation:

$$S_k = G \cdot S_{k-1} + X_k \tag{3.1}$$

[1]The weight w of a binary word is defined as the number of information bits equal to '1', that is the number of information bits differing from the "all zero" word, which is used as a reference for linear codes. For a recursive codes, used in DVB-RCS standard, when the final states are fixed by the encoder, the minimum value for w is 2. For more details see [9] [91], for example.

For Figure 3.1 encoder, vectors S_k and X_k , and matrix G are given by:

$$S_k = \begin{bmatrix} S_{1,k} \\ S_{2,k} \\ S_{3,k} \end{bmatrix}; \qquad X_k = \begin{bmatrix} A_k + B_k \\ B_k \\ B_k \end{bmatrix}; \qquad G = \begin{bmatrix} 1 & 0 & 1 \\ 1 & 0 & 0 \\ 0 & 1 & 0 \end{bmatrix}$$

From Equation (3.1), we can infer:

$$\begin{aligned} S_1 &= G \cdot S_0 + X_1 \\ S_2 &= G \cdot S_1 + X_2 \\ \vdots &\;\vdots\; \vdots \\ S_N &= G \cdot S_{N-1} + X_N \end{aligned} \tag{3.2}$$

Hence, S_N may be expressed as a function of the initial state S_0 and of data feeding the encoder between times 1 and N:

$$S_N = G^N \cdot S_0 + \sum_{p=1}^{N} G^{N-p} \cdot X_p \tag{3.3}$$

It is possible to find a circular state S_c such that $S_c = S_N = S_0$. Its value is derived from Equation (3.3) as:

$$S_c = \left(I + G^N\right)^{-1} \sum_{p=1}^{N} G^{N-p} \cdot X_p \tag{3.4}$$

where I is the identity matrix.

State S_c depends on the sequence of data and exists only if $I + G^N$ is invertible (Note that some G matrices are not suitable). In particular, N cannot be a multiple of the period L of the encoding recursive generator, defined as:

$$G^L = I \tag{3.5}$$

If the encoder starts from state S_c, it comes back to the same state when the encoding of the N data symbols (in Figure 3.1 encoder) is completed. Such an encoding process is called circular because the associated trellis may be viewed as a circle, without any discontinuity on transitions between states.

Determining S_c requires a pre-encoding operation. First, the encoder is initialized in the “all zero” state. Then, the data sequence of length N is encoded once, leading to final state S_N^0. Thus, from Equation (3.3):

$$S_N^0 = \sum_{p=1}^{N} G^{N-p} \cdot X_p \tag{3.6}$$

Combining this result with Equation (3.4), the value of circulation state S_c can be linked to S_N^0 as follows:

$$S_c = \left\langle I + G^N \right\rangle^{-1} S_N^0 \tag{3.7}$$

In a second operation, data are encoded starting from the state S_c calculated from Equation (3.7). The disadvantage of this method is having to encode the sequence twice: once from the "all zero" state and the second time from the state S_c. Nevertheless, in most cases, the double encoding operation can be performed at a frequency much higher than the data rate, so as to reduce the latency effects.

3.1.4 Iterative Decoding Principle for Circular Recursive Codes

Circular codes are well suited to the turbo decoding concept. In fact, the circular code principle may be applied in two slightly different ways, according to whether the code is self-concatenated or not.

Case 1: The code is self-concatenated, that is the second encoding step directly follows on from the first step without intermediate reinitializing of the register state. The circulation state S_c is calculated for the whole sequence of length $2N$. At reception, the decoder performs a decoding of the second sequence of length N.

Case 2: The code is not self-concatenated, that is the encoder is initialized at the beginning of each encoding stage. Two circulation states S_{c_1} and S_{c_2} corresponding to both encoded sequences, are calculated. At reception, the two sequences of length N are decoded separately.

Depending on the case, data encoding is represented by the second circular trellises. Whatever elementary algorithm is used, iterative decoding requires repeated turns around the circular trellis(es), the extrinsic information table being continuously updated during data processing. Iterations naturally follow one after the other without any discontinuity between transitions from state to state.

In the case where the MAP (maximum *a posteriori*) algorithm or the simplified algorithm, Max-Log-MAP algorithm, is applied, decoding the sequence consists of going round the circular trellis anti-clockwise for the backward process, and clockwise for the forward process (Figure 3.2) [90], during which data is decoded and extrinsic information is built. For both processes, probabilities computed at the end of a turn are used as the initial values for the next turn. The number of turns performed around the circular trellis is equal to the number of iterations required by the iterative process. At time k, state S_k can

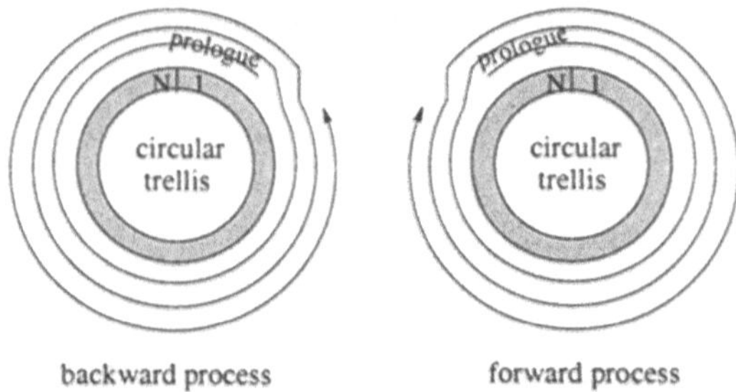

Figure 3.2. Processing a Circular Code by the Backward-forward Algorithm

be represented as,

$$S_k = G^k \cdot S_0 + \sum_{p=1}^{k} G^{k-p} \cdot X_p \tag{3.8}$$

So, knowing state S_k, we can find S_0 using

$$S_0 = \left(I + G^k\right)^{-1} \sum_{p=1}^{k} G^{k-p} \cdot X_p \tag{3.9}$$

where $k = m \cdot v$, m is a small number. This is intended to guide the process towards an initial state which is a good estimate of the circulation state.

3.2. Double-binary CRSC Codes in DVB/RCS Standard

The DVB Committee has approved DVB-RCS standard for Return Channel via Satellite [28]. This is also the ETSI (*European Telecommunications Standards Institute*) standard to provide two-way, full-IP, asymmetric communications via satellite in order to supplement the coverage of Asymmetric Digital Subscriber Line (ADSL) and Cable modem. This standard specifies an air interface allowing a large number of small terminals to send "return" signals to a central gateway and at the same time receive IP data from that hub on the "forward" link in the usual DVB/MPEG2 (Digital Video Broadcasting/Moving Picture Expert Group-2) broadcast format, which places satellite in a favorable position.

In the DVB-RCS standard, the satellite resource on the return link, terminals-to-hub, whose speed can range from 144 Kbps to 2 Mbps, is shared among the terminals transmitting small packets and using MFTDMA (*Multi-Frequency Time Division Multiple Access*) / DAMA (*Demand-assigned multiple access*) techniques.

Since DVB-RCS applications involve the transmission of data using various block sizes and coding rates, the coding scheme has to be very flexible. On the other hand, it has to be able to process data so as to allow the transmission of data bit rates up to 2Mbps. The double-binary CRSC codes are good candidates due to their efficiency in encoding blocks of data, simple puncturing device and interleaver, using the same decoding hardware for every block size/coding rate combination.

3.2.1 System Model

Figure 3.3 shows the system model of the double-binary convolutional turbo code. Coding for channel error protection is applied to traffic and control data, which are transmitted in the types of bursts. In this chapter, the AWGN channel, QPSK modulation and demodulation are used as in the DVB-RCS standard. The complete system model in detail is depicted in Figure 3.12.

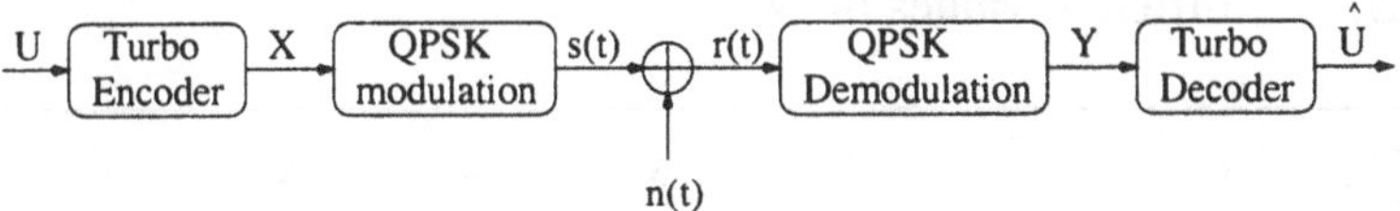

Figure 3.3. System Model of DVB-RCS Standard

3.2.1.1 Encoder Structure. The encoder structure is depicted in Figure 3.4. The data sequence to be encoded, made up of k information bits, feeds the CRSC encoder twice: first, in the natural order of the data (switch in position 1), and next in an interleaved order, given by time permutation function Π (switch in position 2). The encoder is fed by blocks of k bits or N couples ($k = 2 * N\ bits$). N is a multiple of 4 (k is a multiple of 8). The MSB (*Most Significant Bit*) of the first byte after the burst preamble is assigned to A, the next bit to B and so on for the remainder of the burst content.

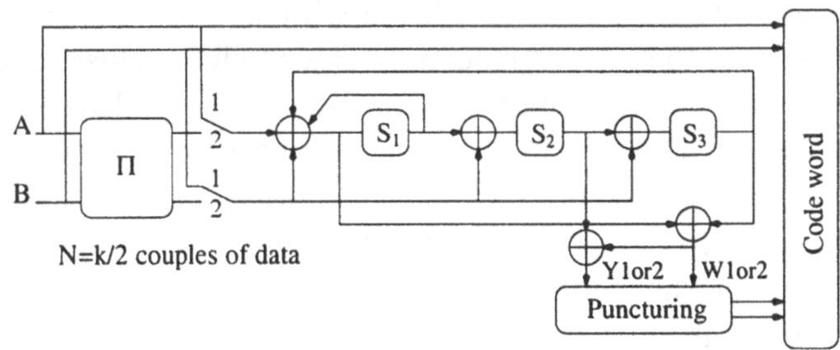

Figure 3.4. Double-binary Circular Recursive Systematic Convolutional Encoder

The polynomials defining the connections are described in octal and symbolic notations as follows:

- for the feedback branch: 15(in octal), equivalently, $1+D+D^3$ (in symbolic notation);

- for the Y parity bits: 13, equivalently, $1 + D^2 + D^3$;

- for the W parity bits: 11, equivalently, $1 + D^3$.

The input A is connected to tap "1" of the shift register and the input B is connected to the taps "1", D and D^2. The state of the encoder is denoted $S(0 \leq S \leq 7)$ with $S = 4 \cdot s_1 + 2 \cdot s_2 + s_3$. Since the value of the circulation state depends on the contents of the sequence to be encoded, determining the circulation state S_c requires a pre-encoding operation.

First, the encoder is initialized in the "all zero" state and fed by the sequence in the natural order with incremental address $i = 0, \ ..., \ N-1$. The data sequence is encoded once, leading to a final state S_N^0. S_c value is then calculated from the expression $S_c = \left(I + G^N\right)^{-1} \cdot S_N^0$. Table 3.1 shows the relationship between S_c and S_N^0 for different values of N.

$S_N^0 \rightarrow$ $N\ mod.7$	0	1	2	3	4	5	6	7
1	$S_c = 0$	$S_c = 6$	$S_c = 4$	$S_c = 2$	$S_c = 7$	$S_c = 1$	$S_c = 3$	$S_c = 5$
2	$S_c = 0$	$S_c = 3$	$S_c = 7$	$S_c = 4$	$S_c = 5$	$S_c = 6$	$S_c = 2$	$S_c = 1$
3	$S_c = 0$	$S_c = 5$	$S_c = 3$	$S_c = 6$	$S_c = 2$	$S_c = 7$	$S_c = 1$	$S_c = 4$
4	$S_c = 0$	$S_c = 4$	$S_c = 1$	$S_c = 5$	$S_c = 6$	$S_c = 2$	$S_c = 7$	$S_c = 3$
5	$S_c = 0$	$S_c = 2$	$S_c = 5$	$S_c = 7$	$S_c = 1$	$S_c = 3$	$S_c = 4$	$S_c = 6$
6	$S_c = 0$	$S_c = 7$	$S_c = 6$	$S_c = 1$	$S_c = 3$	$S_c = 4$	$S_c = 5$	$S_c = 2$

Table 3.1. Circulation State Correspondence Table

Then, the encoder is fed by the same sequence in the natural order with the circulation state S_{c_1}. This first encoding is called C_1 encoding.

Second, the encoder (after initialization) is fed by the interleaved sequence with incremental address $j = 0, \ ..., \ N-1$ with the circulation state S_{c_2} found after pre-encoding in the same manner as in C_1 encoding. This second encoding is called C_2 encoding. The permutation function $i = \prod(j)$ that gives the natural address i of the considered couple, when reading it at place j for the second encoding, is given in subsection 3.2.1.2.

Therefore, to perform a complete encoding operation of the data sequence, two circulation states have to be determined, one for each component encoder, and the sequence has to be encoded four times instead of twice. This is not a real problem, as the encoding operation can be performed at a frequency much higher than the data rate [36].

Figure 3.5 shows the trellis diagram of the above double-binary convolutional turbo encoder.

There are 8 states in the trellis and the numbers shown on the left of each state represent the inputs to the encoder and their corresponding trellis outputs. The numbers from left to right correspond to state transitions from top to bottom exiting from each state.

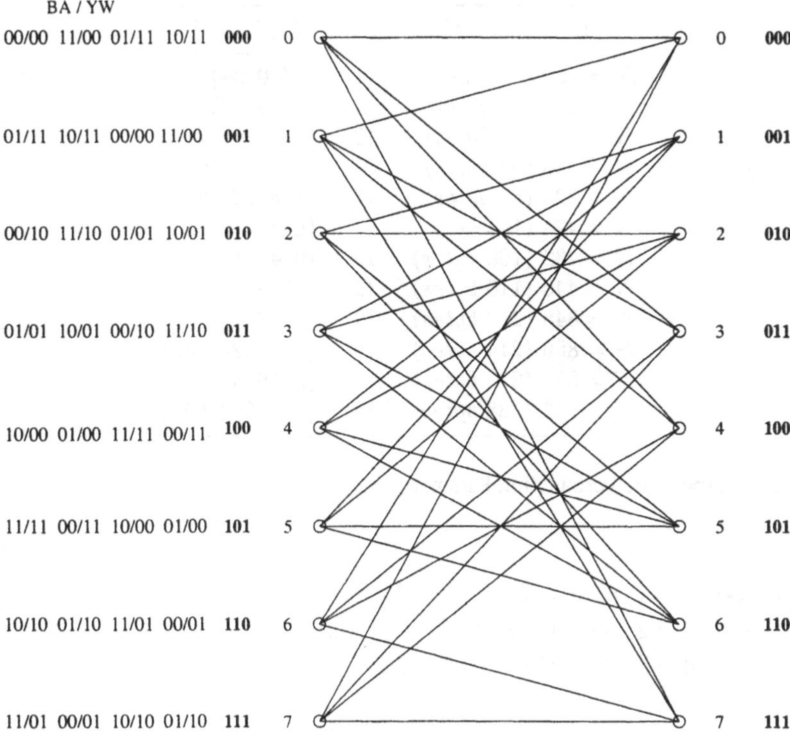

Figure 3.5. Trellis Diagram of CRSC Turbo Code

3.2.1.2 Description of Permutation.

Different permutations (interleavers) can be obtained using generic equations with only a restricted number of parameters. Moreover, a simple puncturing device is sufficient to select the coding rate and, as a result, the same decoding hardware can be used to manage every block size/coding rate combination.

Let N be the number of data couples in each block at the encoder input (each block contains $2N$ data bits). The permutation is done in two levels, the first one inside the couples (**level 1**), and the second one between the couples (**level 2**):

Table 3.2 gives the default permutation parameters P_0, P_1, P_2 and P_3 to be used for different block lengths. These parameters can be updated by the *Time-slot Composition Table* (TCT) (See DVB-RCS standard sub-clause 8.5.5.4 [28]).

level 1

If $j \ mod.\ 2 = 0$ ($j = 0, \ldots, N-1$), let $(A, B) = (B, A)$ (invert the couple)

level 2

- If j mod. 4 = 0, then $P = 0$;

Frame size in couples	P_0	$\{P_1, P_2, P_3\}$
$N = 48$ (12 *bytes*)	11	$\{24, 0, 24\}$
$N = 64$ (16 *bytes*)	7	$\{34, 32, 2\}$
$N = 212$ (53 *bytes*)	13	$\{106, 108, 2\}$
$N = 220$ (55 *bytes*)	23	$\{112, 4, 116\}$
$N = 228$ (57 *bytes*)	17	$\{116, 72, 188\}$
$N = 424$ (106 *bytes*)	11	$\{6, 8, 2\}$
$N = 432$ (108 *bytes*)	13	$\{0, 4, 8\}$
$N = 440$ (110 *bytes*)	13	$\{10, 4, 2\}$
$N = 848$ (212 *bytes*)	19	$\{2, 16, 6\}$
$N = 856$ (214 *bytes*)	19	$\{428, 224, 652\}$
$N = 864$ (216 *bytes*)	19	$\{2, 16, 6\}$
$N = 752$ (188 *bytes*)	19	$\{376, 224, 600\}$

Table 3.2. Turbo Code Permutation Parameters

- If j mod. 4 = 1, then $P = N/2 + P_1$;
- If j mod. 4 = 2, then $P = P_2$;
- If j mod. 4 = 3, then $P = N/2 + P_3$;

$$i = (P_0 * j + P + 1) \ mod. \ N \tag{3.10}$$

The interleaving relations satisfy the odd/even rule, *i.e.*, when j is even, i is odd and vice-versa. This enables the puncturing patterns to be identical for both encodings.

3.2.1.3 Rates and Puncturing Maps. There are seven code rates defined in DVB-RCS standard: R=1/3, 2/5, 1/2, 2/3, 3/4, 4/5, 6/7. These rates are achieved through selectively deleting the parity bits (puncturing). The puncturing patterns of Table 3.3 are applied. These patterns are identical for both codes C_1 and C_2 (deletion is always done in couples). The puncturing rate is indicated to the *Return Channel Satellite Terminals* (RCSTs) via the Time-slot Composition Table (TCT) (See DVB-RCS standard sub-clause 8.5.5.4 [28]).

When the code rate $R \geq 1/2$, all the second parity bits W are deleted. Rates 1/3, 2/5, 1/2, 2/3 and 4/5 are exact, independently of the block size. Rates 3/4 and 6/7 are exact only if N is a multiple of 3. In other cases, the actual rate is very slightly lower than the nominal one.

Depending on the code rate, the length of the encoded block is given in Table 3.4:

3.2.1.4 Order of Transmission and Mapping to QPSK Constellation. Two orders of transmission are allowed:

$$1/3 \quad \begin{matrix} Y \\ W \end{matrix} \begin{bmatrix} 1 & 1 & 1 & 1 & 1 & 1 & 1 & 1 \\ 1 & 1 & 1 & 1 & 1 & 1 & 1 & 1 \end{bmatrix} \qquad 2/5 \quad \begin{matrix} Y \\ W \end{matrix} \begin{bmatrix} 1 & 1 & 1 & 1 & 1 & 1 & 1 & 1 \\ 1 & 0 & 1 & 0 & 1 & 0 & 1 & 0 \end{bmatrix}$$

$$1/2 \quad \begin{matrix} Y \\ W \end{matrix} \begin{bmatrix} 1 & 1 & 1 & 1 & 1 & 1 & 1 & 1 \\ 0 & 0 & 0 & 0 & 0 & 0 & 0 & 0 \end{bmatrix} \qquad 2/3 \quad \begin{matrix} Y \\ W \end{matrix} \begin{bmatrix} 1 & 0 & 1 & 0 & 1 & 0 & 1 & 0 \\ 0 & 0 & 0 & 0 & 0 & 0 & 0 & 0 \end{bmatrix}$$

$$3/4 \quad \begin{matrix} Y \\ W \end{matrix} \begin{bmatrix} 1 & 0 & 0 & 1 & 0 & 0 & 1 & 0 \\ 0 & 0 & 0 & 0 & 0 & 0 & 0 & 0 \end{bmatrix} \qquad 4/5 \quad \begin{matrix} Y \\ W \end{matrix} \begin{bmatrix} 1 & 0 & 0 & 0 & 1 & 0 & 0 & 0 \\ 0 & 0 & 0 & 0 & 0 & 0 & 0 & 0 \end{bmatrix}$$

$$6/7 \quad \begin{matrix} Y \\ W \end{matrix} \begin{bmatrix} 1 & 0 & 0 & 0 & 0 & 0 & 1 & 0 \\ 0 & 0 & 0 & 0 & 0 & 0 & 0 & 0 \end{bmatrix}$$

Table 3.3. Puncturing Patterns for Double-binary Convolutional Turbo Codes. "1" = keep

Encoded block /Code Rate	$2N+M$ *for* $R<1/2$	$N+M$ *for* $R \geq 1/2$
$R=1/3$	$M=N$	
$R=2/5$	$M=N/2$	
$R=1/2$		$M=N$
$R=2/3$		$M=N/2$
		$M=N/3 \quad if\ N\,mod.\,3=0$
$R=3/4$		$M=(N-4)/3+2 \quad if\ N\,mod.\,3=1$
		$M=(N-8)/3+3 \quad if\ N\,mod.\,3=2$
$R=4/5$		$M=N/4$
		$M=N/6 \quad if\ N\,mod.\,3=0$
$R=6/7$		$M=(N-4)/6+1 \quad if\ N\,mod.\,3=1$
		$M=(N-8)/6+2 \quad if\ N\,mod.\,3=2$

Table 3.4. The Length of the Encoded Block

- in the natural order, all couples (A, B) are transmitted first, followed by all couples (Y_1, Y_2) that remain after puncturing and then all couples (W_1, W_2) that remain after puncturing (see Figure 3.6);

- in the reverse order, the couples (Y_1, Y_2) are transmitted first, in their natural order, followed by the couples (W_1, W_2) , if any , and then finally followed by the couples (A, B).

Each couple is mapped to one QPSK constellation point as shown in Figure 3.8. In Figure 3.6, the row with the "A" symbols is mapped on the I channel (C_1 in Figure 3.8). The signal shall be modulated using QPSK, with baseband shaping. Immediately after the preamble insertion, the outputs C_1 and C_2 of the encoder shall be sent without modification to the QPSK bit mapper (see Figure 3.7).

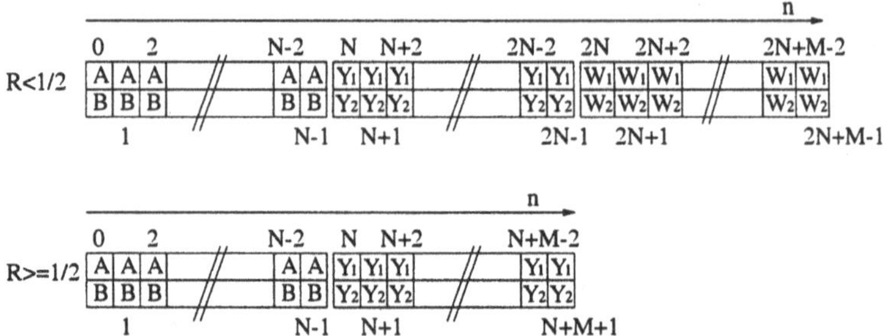

Figure 3.6. Encoded Blocks (Natural Order)

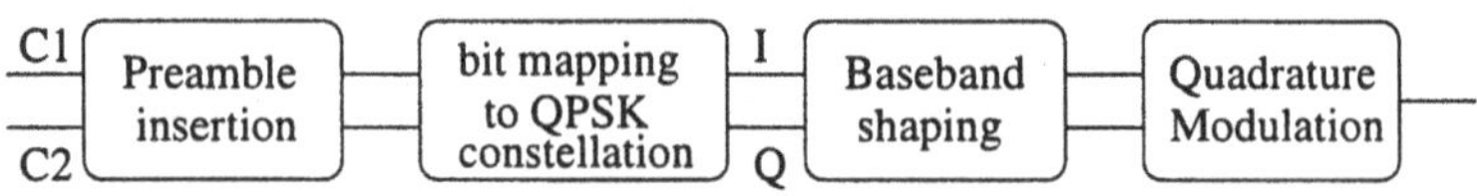

Figure 3.7. Processing after the Encoder

Gray-coded QPSK modulation with absolute mapping (no differential coding) shall be used. Bit mapping in the QPSK constellation shall follow Figure 3.8. If the normalization factor $1/\sqrt{2}$ is applied to the I and Q components, the corresponding average energy per symbol will be 1.

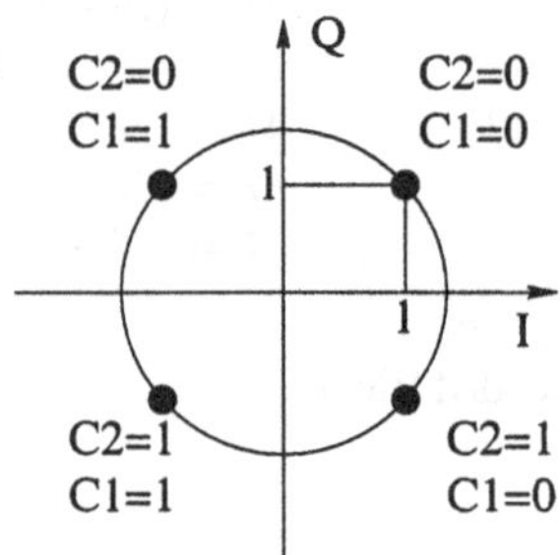

Figure 3.8. Bit Mapping into QPSK Constellation

The output C_1 of the channel coding shall be mapped to the I channel of the modulation. The output C_2 shall be mapped to the Q channel of the modulation.

3.2.2 Decoder Structure

According to the principle of iterative decoding algorithm, the decoder of double-binary CRSC code is designed as shown in Figure 3.9:

The systematic information is the channel value of information symbols $d_k \in \{00, 01, 10, 11\}$. Parity 1 and Parity 2 are the channel value of the outputs

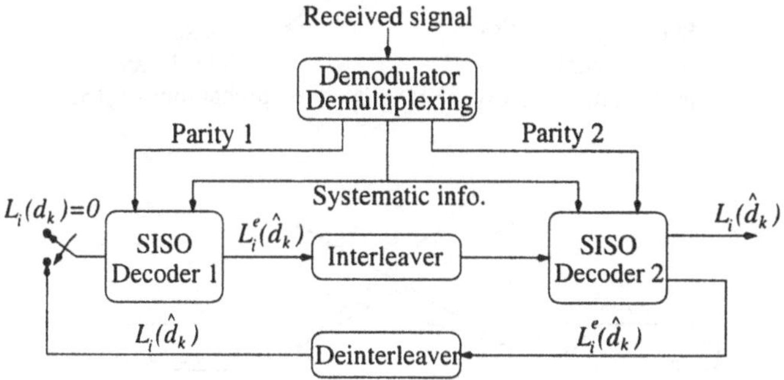

Figure 3.9. Decoder Structure of Non-binary Convolutional Turbo Code

of encoders parity bits. $L_i(\hat{d}_k)$ is the log-likelihood ratio for $i = 1,\ 2,\ 3$ and $L_i^e(\hat{d}_k)$ is the extrinsic information.

3.3. Decoding Procedure of Double-binary Convolutional Turbo Codes

Even though the symbol-by-symbol maximum *a posteriori* (MAP) algorithm is optimal, from an implementation point of view, the component decoding algorithm applied is the Max-Log-MAP algorithm for the complexity/performance compromise. Good convergence, close to the theoretical limits [93]- from 1.0dB to 1.8dB, depending on the coding rate - can be observed, thanks to the double-binary component code.

3.3.1 Decoding Rule for CRSC Codes with a Non-binary Trellis

The trellis of a double-binary feedback convolutional encoder has the structure shown in Figure 3.10. Let S_k be the encoder state at time k. The symbols d_k is associated with the transition from time $k-1$ to time k. The trellis states at level $k-1$ and at level k are indexed by the integer S_{k-1} and S_k, respectively.

The goal of the MAP algorithm is to provide us with

$$\begin{aligned} L_i(d_k) &= \ln \frac{P_r[d_k = i|Observation]}{P_r[d_k = 0|Observation]} \qquad for\ i = 1,2,3 \\ &= \ln \frac{\sum_{d_k=i}^{(S_{k-1},S_k)} p(S_{k-1}, S_k, y_k)}{\sum_{d_k=0}^{(S_{k-1},S_k)} p(S_{k-1}, S_k, y_k)} \end{aligned} \tag{3.11}$$

The index pair $(S_{k-1},\ S_k)$ determines the information symbols (coupled bits in a symbol) d_k and the coded symbols x_k, where d_k is in GF(2^2) with elements {0,1,2,3} from time $k-1$ to time k. The sum of the joint probabilities $p(S_{k-1}, S_k, y_k)$ in the numerator or in the denominator of Equation (3.11) is

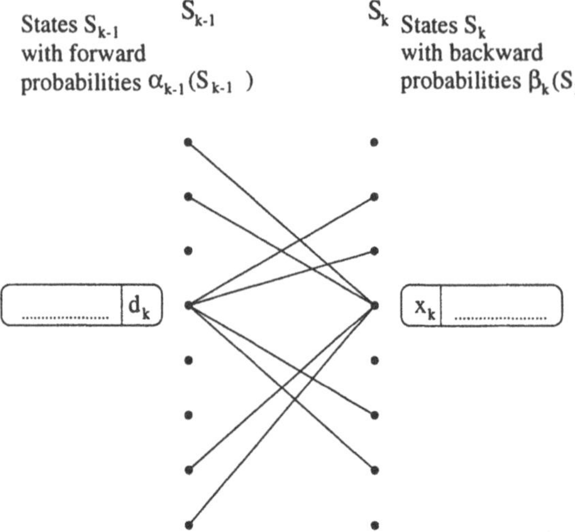

Figure 3.10. Trellis Structure of Double-binary Convolutional Codes with Feedback Encoder

taken over all existing transitions from state S_{k-1} to state S_k labeled with the information bits $d_k = 0, 1, 2, 3$ (that is, $d_k = 00, 01, 10, 11$. We use decimal notation instead of binary for simplicity. Assuming a memoryless transmission channel, the joint probability $p(S_{k-1}, S_k, y_k)$ can be written as the product of three independent probabilities

$$\begin{aligned} p(S_{k-1}, S_k, y_k) &= p(S_{k-1}, y_{j<k}) \cdot p(S_k, y_k|S_{k-1}) \qquad \cdot p(y_{j>k}|S_k) \\ &= \underbrace{p(S_{k-1}, y_{j<k})} \cdot \underbrace{P(S_k|S_{k-1}) \cdot p(y_k|S_{k-1}, S_k)} \cdot \underbrace{p(y_{j>k}|S_k)} \\ &= \alpha_{k-1}(S_{k-1}) \qquad \cdot \gamma_k(S_{k-1}, S_k) \qquad \cdot \beta_k(S_k) \end{aligned} \tag{3.12}$$

Here $y_{j<k}$ denotes the sequence of received symbols y_j from the beginning of the trellis up to time $k-1$ and $y_{j>k}$ is the corresponding sequence from time $k+1$ up to the end of the trellis. The forward recursion of the MAP algorithm yields

$$\alpha_k(S_k) = \sum_{S_{k-1}} \gamma_k(S_{k-1}, S_k) \cdot \alpha_{k-1}(S_{k-1}) \tag{3.13}$$

The backward recursion yields

$$\beta_{k-1}(S_{k-1}) = \sum_{S_k} \gamma_k(S_{k-1}, S_k) \cdot \beta_k(S_k) \tag{3.14}$$

Whenever a transition between S_{k-1} and S_k exist, the branch transition probabilities are given by

$$\gamma_k(S_{k-1}, S_k) = P(S_k|S_{k-1}) \cdot p(y_k|S_{k-1}, S_k) = p(y_k|d_k) \cdot P(d_k) \tag{3.15}$$

Find the natural logarithm of the branch transition probability metrics as

$$\overline{\gamma_k}(S_{k-1}, S_k) = \ln \gamma_k(S_{k-1}, S_k) \tag{3.16}$$

and the natural logarithm of $\alpha_k(S_k)$ and $\beta_k(S_k)$ as

$$\begin{aligned}\overline{\alpha_k}(S_k) &= \ln \alpha_k(S_k) \\ &= \ln \sum_{S_{k-1}} e^{\overline{\gamma_k}(S_{k-1},S_k)} \cdot e^{\overline{\alpha_{k-1}}(S_{k-1})}\end{aligned} \tag{3.17}$$

$$\begin{aligned}\overline{\beta_{k-1}}(S_{k-1}) &= \ln \beta_{k-1}(S_{k-1}) \\ &= \ln \sum_{S_k} e^{\overline{\gamma_k}(S_{k-1},S_k)} \cdot e^{\overline{\beta_k}(S_k)}\end{aligned} \tag{3.18}$$

For clarity we denote the $\gamma_k(S_{k-1}, S_k)$ corresponding to the branch with the symbol $d_k = i$ connecting states S_{k-1} and S_k as $\gamma_k^i(S_{k-1}, S_k)$ in the following equations, for $i = 0, 1, 2, 3$. Hence, the log-likelihood ratios for $i = 1, 2, 3$ are represented by

$$\begin{aligned}L_i(d_k) &= \ln \frac{\sum_{d_k=i}^{(S_{k-1},S_k)} \gamma_k^i(S_{k-1}, S_k) \cdot \alpha_{k-1}(S_{k-1}) \cdot \beta_k(S_k)}{\sum_{d_k=0}^{(S_{k-1},S_k)} \gamma_k^0(S_{k-1}, S_k) \cdot \alpha_{k-1}(S_{k-1}) \cdot \beta_k(S_k)} \\ &= \ln \frac{\sum_{d_k=i}^{(S_{k-1},S_k)} e^{\overline{\gamma_k^i}(S_{k-1},S_k)} \cdot e^{\overline{\alpha_{k-1}}(S_{k-1})} \cdot e^{\overline{\beta_k}(S_k)}}{\sum_{d_k=0}^{(S_{k-1},S_k)} e^{\overline{\gamma_k^0}(S_{k-1},S_k)} \cdot e^{\overline{\alpha_{k-1}}(S_{k-1})} \cdot e^{\overline{\beta_k}(S_k)}}\end{aligned} \tag{3.19}$$

3.3.2 Simplified Max-Log-MAP Algorithm for Double-binary Convolutional Turbo Code

In order to reduce the computational complexity, to increase throughput, or to reduce the power consumption, we consider the sub-optimal Max-Log-MAP algorithm for the non-binary convolutional turbo codes. Extrinsic information coupling (for the feedback) is performed according to Hagenauer [7].

First, according to the decoding rule and Equation (3.15) and (3.16), find the logarithm of the branch transition probability as:

$$\overline{\gamma_k^i}(S_{k-1}, S_k) = \ln \gamma_k^i(S_{k-1}, S_k) = \ln p(y_k|d_k) \cdot P(d_k) \tag{3.20}$$

The distribution of the received parity and systematic symbols are given by

$$\begin{aligned}& p[y_k|d_k = i] \qquad\qquad for\ i = 0, 1, 2, 3 \\ =\ & p[y_k^s|x_k^s(i)] \cdot p[y_k^p|x_k^p(i, S_{k-1}, S_k)] \\ =\ & \frac{1}{\pi \cdot N_0} exp\{-\frac{E_s}{N_0}[(y_k^{s,I} - x_k^{s,I}(i))^2 + (y_k^{s,Q} - x_k^{s,Q}(i))^2]\}\end{aligned}$$

$$\begin{aligned}
&\cdot\frac{1}{\pi\cdot N_0}exp\{-\frac{E_s}{N_0}[(y_k^{p,I}-x_k^{p,I}(i,S_{k-1},S_k))^2+(y_k^{p,I}-x_k^{p,Q}(i,S_{k-1},S_k))^2]\}\\
=\;& B_k\cdot exp\{\frac{1}{2}L_c\cdot[y_k^{s,I}\cdot x_k^{s,I}(i)+y_k^{s,Q}\cdot x_k^{s,Q}(i)]\}\\
&\cdot exp\{\frac{1}{2}L_c\cdot[y_k^{p,I}\cdot x_k^{p,I}(i,S_{k-1},S_k)+y_k^{p,I}\cdot x_k^{p,Q}(i,S_{k-1},S_k)]\} \qquad (3.21)
\end{aligned}$$

where y_k^s, y_k^p represent the received systematic and parity symbols, and $y_k^{s,I}$, $y_k^{s,Q}$, $y_k^{p,I}$, $y_k^{p,Q}$ represent the received bit values transmitted through the I and Q channel, respectively; $x_k^s(i)$, $x_k^p(i,S_{k-1},S_k)$ represent the systematic and parity symbols for $i=0,1,2,3$, and $x_k^{s,I}(i)$, $x_k^{s,Q}(i)$, $x_k^{p,I}(i,S_{k-1},S_k)$, $x_k^{p,Q}(i,S_{k-1},S_k)$ represent the bits of codeword mapped to QPSK constellation, respectively.

Here,

$$\begin{aligned}
B_k \;=\;& (\frac{1}{\pi\cdot N_0})^2 exp\{-\frac{E_s}{N_0}[(y_k^{s,I})^2+(x_k^{s,I}(i))^2+(y_k^{s,Q})^2+(x_k^{s,Q}(i))^2\\
&+(y_k^{p,I})^2+(x_k^{p,I}(i,S_{k-1},S_k)^2+(y_k^{p,I})^2+(x_k^{p,Q}(i,S_{k-1},S_k))^2]\} \qquad (3.22)
\end{aligned}$$

Hence,

$$\begin{aligned}
&\overline{\gamma_k}(S_{k-1},S_k)\\
=\;& \ln\gamma_k(S_{k-1},S_k)\\
=\;& \ln P(y_k|d_k)\cdot P(d_k)\\
=\;& \frac{1}{2}L_c\cdot[y_k^{s,I}\cdot x_k^{s,I}(i)+y_k^{s,Q}\cdot x_k^{s,Q}(i)]+\ln P(d_k)+K\\
&+\frac{1}{2}L_c\cdot[y_k^{p,I}\cdot x_k^{p,I}(i,S_{k-1},S_k)+y_k^{p,I}\cdot x_k^{p,Q}(i,S_{k-1},S_k)] \qquad (3.23)
\end{aligned}$$

where the constant K includes the constants and common terms that are cancelled in comparisons at later stages.

Next, compute $\alpha_k(S_k)$ and $\beta_k(S_k)$ as

$$\begin{aligned}
\alpha_k(s_k) \;&=\; \sum_{S_{k-1}}\gamma_k(S_{k-1},S_k)\cdot\alpha_{k-1}(S_{k-1})\\
&=\; \sum_{S_{k-1}} e^{\overline{\gamma_k}(S_{k-1},S_k)}\cdot e^{\overline{\alpha_{k-1}}(S_{k-1})} \qquad (3.24)
\end{aligned}$$

and then take max-function,

$$\begin{aligned}
\overline{\alpha_k}(S_k) \;&=\; \ln\sum_{S_{k-1}} e^{\overline{\gamma_k}(S_{k-1},S_k)}\cdot e^{\overline{\alpha_{k-1}}(S_{k-1})}\\
&\approx\; \max_{S_{k-1}}[\overline{\gamma_k}(S_{k-1},S_k)+\overline{\alpha_{k-1}}(S_{k-1}) \qquad (3.25)
\end{aligned}$$

Similarly

$$\begin{aligned}\overline{\beta_{k-1}}(S_{k-1}) &= \ln\sum_{S_k} e^{\overline{\gamma_k}(S_{k-1},S_k)} \cdot e^{\overline{\beta_k}(S_k)} \\ &\approx \max_{S_k}\{[\overline{\gamma_k}(S_{k-1},S_k)+\overline{\beta_k}(S_k)]\end{aligned} \tag{3.26}$$

For iterative decoding of circular trellis, Tail-biting is

$$\begin{aligned}\overline{\alpha_0}(S_0) &= \overline{\alpha_N}(S_N) \quad for\ \forall S_0 \\ \overline{\beta_N}(S_N) &= \overline{\beta_0}(S_0) \quad for\ \forall s_N\end{aligned} \tag{3.27}$$

Therefore, computing the log-likelihood ratios follows the Equation (3.19) and takes max(·) function as

$$\begin{aligned}L_i(\hat{d}_k) \approx &\max_{(s_{k-1},s_k)}[\overline{\gamma_k^i}(S_{k-1},S_k)+\overline{\alpha_{k-1}}(S_{k-1})+\overline{\beta_k}(S_k)] \\ &-\max_{(s_{k-1},s_k)}[\overline{\gamma_k^0}(S_{k-1},S_k)+\overline{\alpha_{k-1}}(S_{k-1})+\overline{\beta_k}(S_k)]\end{aligned} \tag{3.28}$$

Moreover, to separate the Log-likelihood ratios into intrinsic, systematic and extrinsic information, define:

$$\begin{aligned}\gamma_k^{i(e)}(S_{k-1},S_k) &= exp\{\frac{1}{2}L_c\cdot[y_k^{p,I}\cdot x_k^{p,I}(i,S_{k-1},S_k) \\ &\quad +y_k^{p,I}\cdot x_k^{p,Q}(i,S_{k-1},S_k)]\}\cdot A_k\end{aligned} \tag{3.29}$$

here, the constant

$$\begin{aligned}A_k &= \frac{1}{\pi\cdot N_0}exp\{-\frac{E_s}{N_0}[(y_k^{p,I})^2+(x_k^{p,I}(i,S_{k-1},S_k))^2 \\ &\quad +(y_k^{p,I})^2+(x_k^{p,Q}(i,S_{k-1},S_k))^2]\}\end{aligned} \tag{3.30}$$

Hence, the logarithm of the branch transition operation reduce to the expression with

$$\begin{aligned}&\overline{\gamma_k^{i(e)}}(S_{k-1},S_k) \\ &= \ln\gamma_k^{i(e)}(S_{k-1},S_k) \\ &= \tfrac{1}{2}L_c\cdot[y_k^{p,I}\cdot x_k^{p,I}(i,S_{k-1},S_k)+y_k^{p,I}\cdot x_k^{p,Q}(i,S_{k-1},S_k)]+K'\end{aligned} \tag{3.31}$$

where the constant K' includes the constants and common terms that are cancelled in comparisons in later stages.

In another way, find the Log-likelihood ratios as

$$L_i(\hat{d}_k) = \ln\frac{\sum_{d_k=i}^{(S_{k-1},S_k)}\gamma_k^i(S_{k-1},S_k)\cdot\alpha_{k-1}(S_{k-1})\cdot\beta_k(S_k)}{\sum_{d_k=0}^{(S_{k-1},S_k)}\gamma_k^0(S_{k-1},S_k)\cdot\alpha_{k-1}(S_{k-1})\cdot\beta_k(S_k)} \quad for\ i=1,2,3$$

$$
\begin{aligned}
&= \ln \frac{exp\{\frac{1}{2}L_c \cdot [y_k^{s,I} \cdot x_k^{s,I}(i) + y_k^{s,Q} \cdot x_k^{s,Q}(i)]\} \cdot P[d_k = i]}{exp\{\frac{1}{2}L_c \cdot [y_k^{s,I} \cdot x_k^{s,I}(0) + y_k^{s,Q} \cdot x_k^{s,Q}(0)]\} \cdot P[d_k = 00]} \\
&\quad + \ln \frac{\sum_{d_k=i}^{(S_{k-1},S_k)} \gamma_k^{i(e)}(S_{k-1}, S_k) \cdot \alpha_{k-1}(S_{k-1}) \cdot \beta_k(S_k)}{\sum_{d_k=0}^{(S_{k-1},S_k)} \gamma_k^{0(e)}(S_{k-1}, S_k) \cdot \alpha_{k-1}(S_{k-1}) \cdot \beta_k(S_k)} \\
&= \{\frac{1}{2}L_c \cdot [y_k^{s,I} \cdot x_k^{s,I}(i) + y_k^{s,Q} \cdot x_k^{s,Q}(i)] - \frac{1}{2}L_c \cdot [y_k^{s,I} \cdot x_k^{s,I}(0) + y_k^{s,Q} \cdot x_k^{s,Q}(0)]\} \\
&\quad + \underbrace{\ln \frac{P[d_k = i]}{P[d_k = 0]}}_{L_i(d_k)} + \underbrace{\ln \frac{\sum_{d_k=i}^{(S_{k-1},S_k)} \gamma_k^{i(e)}(S_{k-1}, S_k) \cdot \alpha_{k-1}(S_{k-1}) \cdot \beta_k(S_k)}{\sum_{d_k=0}^{(S_{k-1},S_k)} \gamma_k^{0(e)}(S_{k-1}, S_k) \cdot \alpha_{k-1}(S_{k-1}) \cdot \beta_k(S_k)}}_{L_i^e(\hat{d}_k)}
\end{aligned} \tag{3.32}
$$

So the extrinsic information can be calculated as

$$
L_i^e(\hat{d}_k) = \ln \frac{\sum_{d_k=i}^{(S_{k-1},S_k)} \gamma_k^{i(e)}(S_{k-1}, S_k) \cdot \alpha_{k-1}(S_{k-1}) \cdot \beta_k(S_k)}{\sum_{d_k=0}^{(S_{k-1},S_k)} \gamma_k^{0(e)}(S_{k-1}, S_k) \cdot \alpha_{k-1}(S_{k-1}) \cdot \beta_k(S_k)} \tag{3.33}
$$

$$
\begin{aligned}
L_i^e(\hat{d}_k) &= L_i(\hat{d}_k) - \frac{1}{2}L_c \cdot [y_k^{s,I} \cdot x_k^{s,I}(i) + y_k^{s,Q} \cdot x_k^{s,Q}(i)] \\
&\quad + \frac{1}{2}L_c \cdot [y_k^{s,I} \cdot x_k^{s,I}(0) + y_k^{s,Q} \cdot x_k^{s,Q}(0)] - \ln \frac{P[d_k = i]}{P[d_k = 0]}
\end{aligned} \tag{3.34}
$$

Compute symbol probabilities for the next decoder for $i = 1, 2, 3$ from previous decoder as:

$$
L_i(d_k) = L_i^e(\hat{d}_k) = \ln \frac{P[d_k = i]}{P[d_k = 0]} \tag{3.35}
$$

Since

$$
\begin{cases}
L_0^e(\hat{d}_k) = \ln \frac{P[d_k=00]}{P[d_k=00]} = \ln 1 = 0 \\
L_1^e(\hat{d}_k) = \ln \frac{P[d_k=01]}{P[d_k=00]} \\
L_2^e(\hat{d}_k) = \ln \frac{P[d_k=10]}{P[d_k=00]} \\
L_3^e(\hat{d}_k) = \ln \frac{P[d_k=11]}{P[d_k=00]}
\end{cases} \tag{3.36}
$$

Then

$$\begin{cases} P[d_k = 01] &= e^{L_1^e(\hat{d}_k)} \cdot P[d_k = 00] \\ P[d_k = 10] &= e^{L_2^e(\hat{d}_k)} \cdot P[d_k = 00] \\ P[d_k = 11] &= e^{L_3^e(\hat{d}_k)} \cdot P[d_k = 00] \end{cases} \tag{3.37}$$

and

$$P[d_k = 00] + P[d_k = 01] + P[d_k = 10] + P[d_k = 11] = 1 \tag{3.38}$$

Hence

$$\begin{cases} P[d_k = 00] &= \frac{1}{1+e^{L_1^e(\hat{d}_k)}+e^{L_2^e(\hat{d}_k)}+e^{L_3^e(\hat{d}_k)}} \\ P[d_k = 01] &= \frac{e^{L_1^e(\hat{d}_k)}}{1+e^{L_1^e(\hat{d}_k)}+e^{L_2^e(\hat{d}_k)}+e^{L_3^e(\hat{d}_k)}} \\ P[d_k = 10] &= \frac{e^{L_2^e(\hat{d}_k)}}{1+e^{L_1^e(\hat{d}_k)}+e^{L_2^e(\hat{d}_k)}+e^{L_3^e(\hat{d}_k)}} \\ P[d_k = 11] &= \frac{e^{L_3^e(\hat{d}_k)}}{1+e^{L_1^e(\hat{d}_k)}+e^{L_2^e(\hat{d}_k)}+e^{L_3^e(\hat{d}_k)}} \end{cases} \tag{3.39}$$

Using max-function

$$\begin{cases} \ln P[d_k = 00] &= -\max[0, L_1^e(\hat{d}_k), L_2^e(\hat{d}_k), L_3^e(\hat{d}_k)] \\ \ln P[d_k = 01] &= L_1^e(\hat{d}_k) - \max[0, L_1^e(\hat{d}_k), L_2^e(\hat{d}_k), L_3^e(\hat{d}_k)] \\ \ln P[d_k = 10] &= L_2^e(\hat{d}_k) - \max[0, L_1^e(\hat{d}_k), L_2^e(\hat{d}_k), L_3^e(\hat{d}_k)] \\ \ln P[d_k = 11] &= L_3^e(\hat{d}_k) - \max[0, L_1^e(\hat{d}_k), L_2^e(\hat{d}_k), L_3^e(\hat{d}_k)] \end{cases} \tag{3.40}$$

3.3.3 Initialization and the Final Decision

Assuming equally likely information symbols: we do not have any *a priori* information available for the first iteration, we initialize

$$\begin{cases} L_0(d_k = 00) &= 0 \\ L_1(d_k = 01) &= 0 \\ L_2(d_k = 10) &= 0 \\ L_3(d_k = 11) &= 0 \end{cases} \tag{3.41}$$

and according to Equation (3.39), we have

$$P_0[d_k = 00] = \frac{1}{1 + e^{L_1(d_k)} + e^{L_2(d_k)} + e^{L_3(d_k)}} = \frac{1}{4}$$

$$
\begin{aligned}
P_0[d_k = 01] &= \frac{1}{4} \\
P_0[d_k = 10] &= \frac{1}{4} \\
P_0[d_k = 11] &= \frac{1}{4}
\end{aligned}
$$

Using max-function:

$$
\begin{aligned}
\ln P_0[d_k = 00] &= -max(0, L_1, L_2, L_3) = 0 \\
\ln P_1[d_k = 01] &= L_1 - max(0, L_1, L_2, L_3) = 0 \\
\ln P_2[d_k = 10] &= L_2 - max(0, L_1, L_2, L_3) = 0 \\
\ln P_3[d_k = 11] &= L_3 - max(0, L_1, L_2, L_3) = 0
\end{aligned}
\tag{3.42}
$$

Similarly, from the assumption of equally likely symbols, we have,

$$
\begin{aligned}
\alpha_0(s_0) &= 1 \quad for\ \forall s_0 \\
\beta_N(s_N) &= 1 \quad for\ \forall s_N
\end{aligned}
\tag{3.43}
$$

or

$$
\begin{aligned}
\overline{\alpha_0}(S_0) &= \ln \alpha_0(s_0) &= 0 \quad for\ \forall s_0 \\
\overline{\beta_N}(S_N) &= \ln \beta_N(s_N) &= 0 \quad for\ \forall s_N
\end{aligned}
\tag{3.44}
$$

and we initialize $\overline{\gamma_0^i} = 0$.

The reliability value of the channel

$$
L_c = 4a \cdot \frac{E_s}{N_0} = 4a \cdot R_c \cdot \frac{E_b}{N_0}
\tag{3.45}
$$

where R_c is the code rate.

After several decoding iterations, the decisions are made according to:

$$
\hat{d}_k = \begin{cases}
01 & if\ L(\hat{d}_k) = L_1(\hat{d}_k)\ and\ L_1(\hat{d}_k) > 0 \\
10 & if\ L(\hat{d}_k) = L_2(\hat{d}_k)\ and\ L_2(\hat{d}_k) > 0 \\
11 & if\ L(\hat{d}_k) = L_3(\hat{d}_k)\ and\ L_3(\hat{d}_k) > 0 \\
00 & else
\end{cases}
\tag{3.46}
$$

where

$$
L(\hat{d}_k) = \max(L_1(\hat{d}_k), L_2(\hat{d}_k), L_3(\hat{d}_k))
\tag{3.47}
$$

3.3.4 Simulation Results

Table 3.5 gives some examples of the DVB-RCS turbo code performance observed over a Gaussian channel at *Frame Error Rate* (FER)($FER = 10^{-4}$)[2],

[2]Here, we have 100 bit-error events for all simulations in this chapter.

compared to the theoretical limits [93] and the simulation results reported in [36]. The results in the last column were obtained using the program in the CD-ROM.

Code Rate	*Theoretical limits*	*In [36]*	*Our results*
1/2	1.3 dB	2.3 dB	2.3 dB
2/3	2.2 dB	3.3 dB	3.3 dB
3/4	2.6 dB	3.9 dB	4.05 dB
4/5	3.1 dB	4.6 dB	4.8 dB
6/7	3.8 dB	5.2 dB	5.6 dB

Table 3.5. $E_b/N_0(dB)$ *at* $FER = 10^{-4}$, 8-iteration, Simulation over AWGN Channel with Max-Log-MAP Algorithm. ATM Cells, 53 bytes.

Figure 3.11 exhibits the performance for a block size of 53 bytes with the simplified Max-Log-MAP algorithm. So far, no error floor has been observed. In [36], for FER down to 10^{-8} (equivalent to $BER = 10^{-10} - 10^{-11}$), the measurements show the absence of error floor.

3.4. Summary

Since convolutional turbo codes are very flexible codes, easily adaptable to a large range of data block sizes and coding rates, they have been adopted in the DVB standard for Return Channel via Satellite (DVB-RCS). We followed the specifications of turbo coding/decoding in that standard, for twelve block sizes and seven coding rates, and presented the simulation results, in particular for the transmission of ATM cells in AWGN channel, show the performance of the coding scheme chosen. Moreover, the iterative decoding procedure and simplified iterative decoding algorithm for double-binary convolutional turbo code was presented.

In DVB-RCS standard, the substitution of the binary codes by the double-binary codes has a direct incidence on the erroneous paths in the trellises, which leads to a lowered path error density and reduces the correlation effects in the decoding process. This leads to a performance better than that of binary turbo codes for equivalent implementation complexity. Circular coding is a kind of "tail-biting" technique that avoids reducing the code rate and increasing the transmission bandwidth. Non-uniform interleaving is applied to avoid many error patterns due to adopting double-binary CRSC codes. The influence of puncturing and suboptimal decoding algorithm, Max-Log-MAP algorithm, are less significant with double-binary turbo codes than with binary turbo codes. Using double-binary codes, the latency of the decoder is halved. Therefore,

Bit Error Rate (BER) for seven code rate

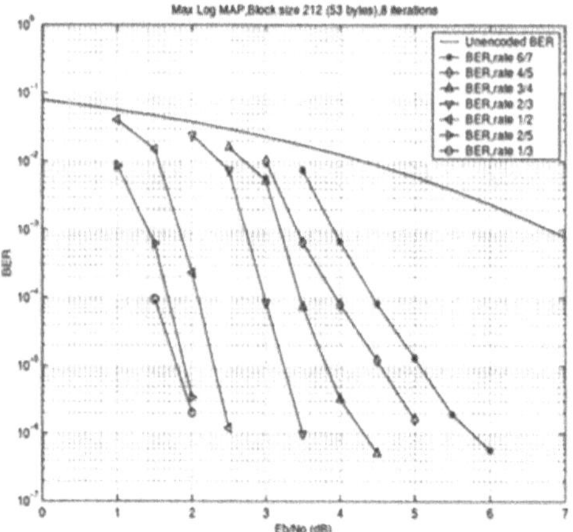

Frame Error Rate (FER) for seven code rate

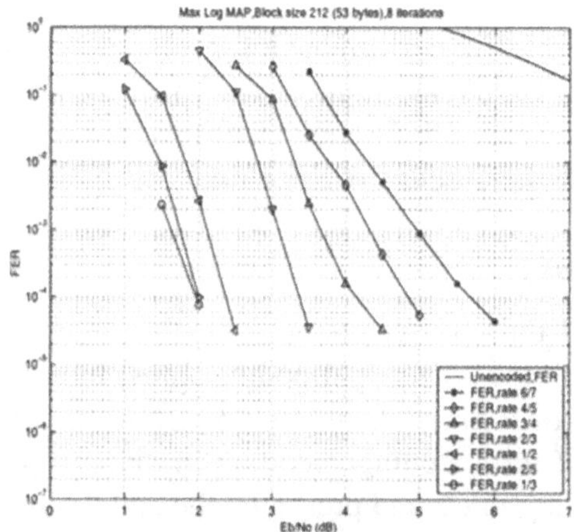

Figure 3.11. Bit Error Rate and Frame Error Rate for Seven Code Rates.

double-binary CRSC code could be easily adopted for many applications, for various block sizes and code rates while retaining excellent coding gains.

In conclusion, double-binary CRSC code that was proposed for DVB-RCS applications is powerful, very flexible and can be implemented with reasonable complexity. Moreover, double-binary CRSC codes are compatible with other techniques applied to error floor optimization. The system model for the whole encoding/decoding procedure is shown in Figure 3.12.

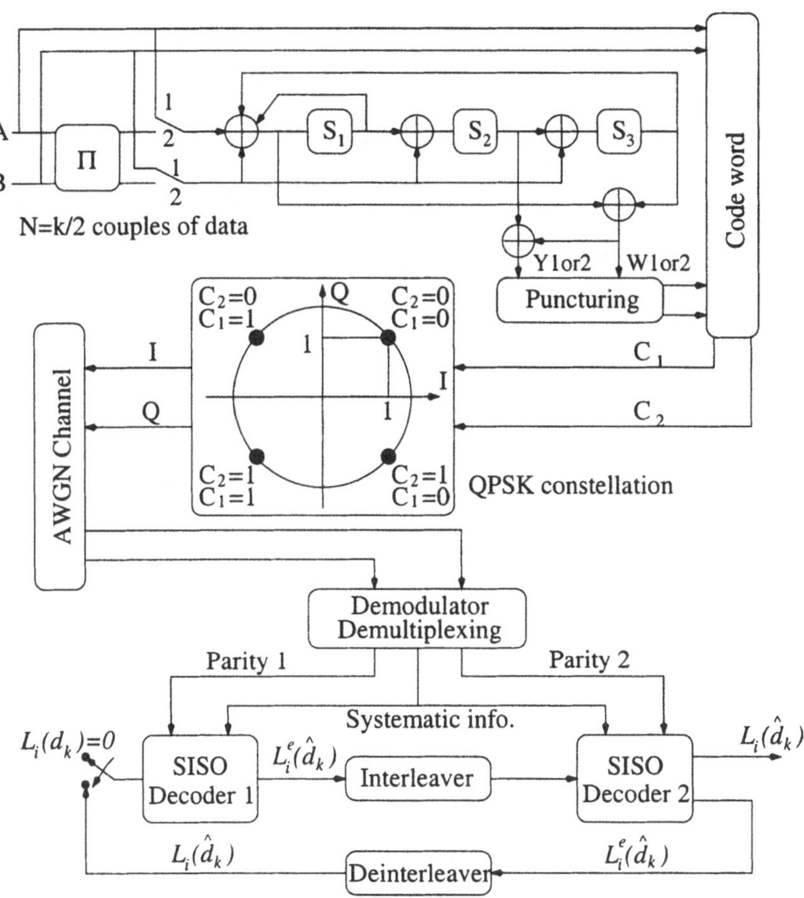

Figure 3.12. Block Diagram of the DVB-RCS Transmition Shcheme

Chapter 4

SPECTRALLY EFFICIENT NON-BINARY TURBO CODES: BEYOND DVB/RCS

Power and bandwidth are limited resources in modern communications systems. Efficient exploitation of these resources will invariably involve an increase in the complexity of a communication system. If the signal set dimensionality per information bit is unchanged, the spectral efficiency remains unchanged. Even though double-binary CRSC codes have an excellent performance, they are limited by the QPSK modulation to a bandwidth efficiency of less than 2bits/s/Hz, as well as the limit on puncturing. There are only limited number of parity bits to be punctured to achieve higher bandwidth efficiency. For example, the number of parity bits left in each encoder for the code rate $6/7$ is only $1/12$ of the information bits.

In this chapter, the design of a triple-binary CRSC code [110] is presented. This code is intended for being used with 8PSK modulation. The turbo encoder design involves the component encoder design, the interleaver design and the puncturer design. Certain special conditions need to be met at the encoder and the iterative decoder need to be adapted to symbol-by-symbol decoding.

4.1. Design of Triple-binary Codes for 8PSK Modulation

Using double-binary codes as component codes represents a simple means to reduce the correlation effects that have a direct incidence on the erroneous paths in the trellises [89]. The use of double-binary turbo codes lead to a lowered path error density and reduces the correlation effects in the decoding process [89]. This leads to a better performance so that an 8-state double-binary turbo code performs better than a 16-state binary turbo code. The degradation resulting from puncturing and using a simplified version of the MAP algorithm is also less significant in the case of double-binary codes [36]. Moreover, from an implementation point of view, the bit rate at the decoder output is twice that of a binary decoder processing for the same number of iterations, with the same circuit clock frequency and with an equivalent complexity per decoded bit. Thus, given the data block size, the latency of the decoder is divided by 2 compared to the binary case because the size of the permutation matrix is halved [36].

Double-binary convolutional turbo codes, thanks to their advantages, have been adopted in the DVB standard. However, they are suitable only for combination with QPSK modulation, and the iterative decoding will be handicapped by the symbol values of the channel output is used with 8PSK modulation. The symbol-by-symbol MAP algorithm and the puncturing map for double-binary codes do not work for double-binary codes combined with 8PSK mapping.

Motivated by the above considerations, the triple-binary codes are designed to be used with 8PSK modulation. The encoder structure is PCCC. The component codes are still CRSC codes. Therefore, there is no need for the addition of the tail bits, and there is no degradation of the spectral efficiency. Non-binary turbo codes are also discussed in [17] and [118].

The decoding principle is the same as that for the double-binary codes. The simplified Max-Log-MAP algorithm applied to double-binary codes is modified to a symbol-by-symbol. Higher decoder speeds are achieved thanks to the use of higher rate convolutional codes in the code construction.

4.2. System Model

Triple-binary code still has the features of the CRSC codes, which avoids the degradation of the spectral efficiency. The system model is similar to DVB-RCS standard and is exhibited in Figure 4.1. The data sequence to be encoded, made up of k information bits, feeds the CRSC encoder twice: first, in the natural order of the data (switch in position 1), and next in an interleaved order, given by the time permutation function Π (switch in position 2).

4.2.1 Constituent Encoder

What is crucial to the practical suitability of turbo codes is the fact that they can be decoded iteratively with good performance. However, the resulting iterative decoder is restricted by the signal mapper of 8PSK constellation, therefore, certain special conditions need to be met at the encoder. The encoder is fed by blocks of k bits or N triplets ($k = 3 * N\ bits$). To have an integer number of bytes in each packet, N needs to be an integer multiple of 8. As a result, the information length k is a multiple of 24 since 3 and 8 are mutually prime. It uses a triple-binary CRSC code shown in Figure 4.2, with three parallel input bits, three parallel systematic bits and three parallel parity bits. The 3-bit input and output, makes this code convenient for coupling with 8PSK mapping with no resulting numberical complication in the iterative decoding procedure [111]. This also allows for flexible puncturing in order to obtain higher code rates.

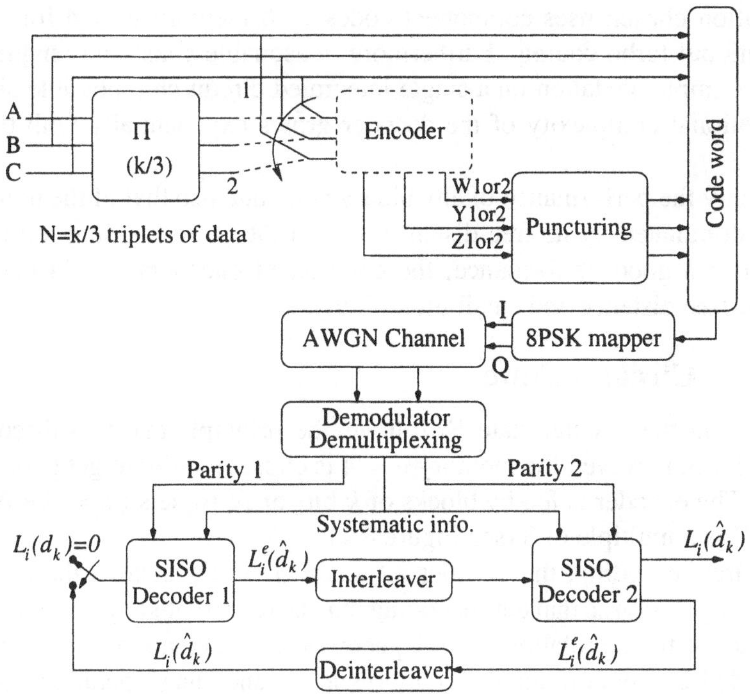

Figure 4.1. System Model of Triple-binary Code Combined 8PSK Modulation

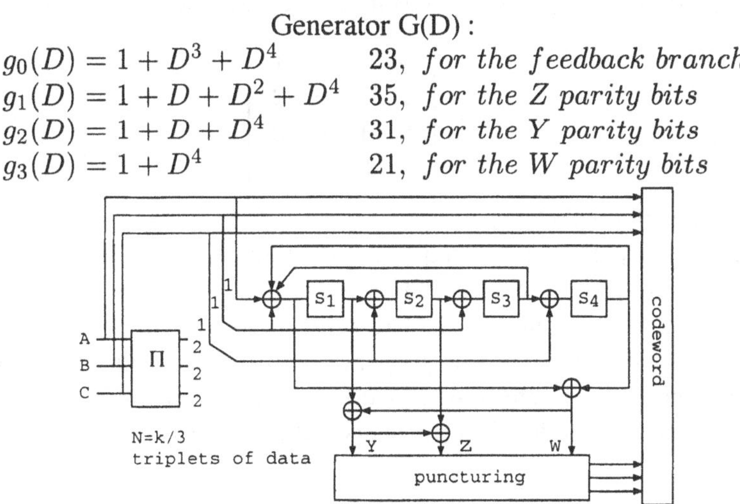

Figure 4.2. Encoder Structure with Generator G(D)

Since the constituent codes with small constraint lengths ensure convergence at very low signal to noise ratios and the correlation effects are minimized, the solution chosen uses component codes with memory $v = 4$ for efficient convolutional turbo coding. Furthermore, reasonable constraint lengths make hardware implementation on a single integrated circuit chip possible since the computational complexity of the decoder grows exponentially with the code memory.

Not only the performance of any binary code but also that of the non-binary code is dominated by its free distance d_f and the corresponding multiplicity. To achieve a good performance, the component encoders should have large effective free distance and small multiplicity.

4.2.2 Circular State

Determining the circular state S_c follows the principle that was discussed in section 3.1.3, however, the memory $v = 4$ is chose in order to get good performance. The encoder is fed by blocks of k bits or N triplets ($k = 3 * N$ bits), where N is a multiple of 8 (see Figure 4.2).

For circular coding, the encoder retrieves the initial state at the end of the encoding operation so that data encoding may be represented by a circular trellis. The value of the circulation state depends on the contents of the sequence to be encoded and determining S_c requires a pre-encoding operation: first, the encoder is initialized in the "all zero" state. The data sequence is encoded once, leading to a final state S_N^0. Then, S_c is calculated from the expression $S_c = \left(I + G^N\right)^{-1} \cdot S_N^0$ as discussed in Chapter 2. The matrix G is given by:

$$G = \begin{bmatrix} 0 & 0 & 1 & 1 \\ 1 & 0 & 0 & 0 \\ 0 & 1 & 0 & 0 \\ 0 & 0 & 1 & 0 \end{bmatrix}$$

The state of the encoder is denoted by $S \, (0 \leq S \leq 15)$ with $S = 8 \cdot s_1 + 4 \cdot s_2 + 2 \cdot s_3 + s_4$. For a given sequence of length N, use Table 4.1 to find S_c. In Table 4.1, we only show three different frame sizes, $N = 152$ $(57 \; bytes)$, $N = 752$ $(282 \; bytes)$ and $N = 224$ $(84 \; bytes)$ as example.

To perform a complete encoding operation of the data sequence, two circulation states have to be determined, one for each component encoder, and the sequence has to be encoded four times instead of twice as described in Chapter 3.

4.2.3 Description of the Turbo Code Permutation

For double-binary CRSC codes, non-uniform interleaving can be implemented by introducing local disorder into the data couples. For example, (A, B) become (B, A), or (B, A+B), *etc.* However, for triple-binary CRSC codes, non-uniformity makes the iterative decoding very difficult and complex because of

$N\ mod.15 \rightarrow$ S_N^0	$N = 152$ $N = 752$	$N = 224$
0	0	0
1	11	15
2	13	1
3	6	14
4	10	3
5	1	12
6	7	2
7	12	13
8	5	7
9	14	8
10	8	6
11	3	9
12	15	4
13	4	11
14	2	5
15	9	10

Table 4.1. Circulation State Correspondence Table for Triple-binary Codes

8PSK constellation mapping. Therefore, we only use inter-symbol interleaving. To achieve the different permutations that govern the performance of PCCC, at low error rates, we use generic equations with only a restricted number of parameters. The permutation parameters are still denoted as P_0, P_1, P_2 and P_3 with formula $i = (P_0 * j + P + 1) mod.\, N$.

Let N be the number of data triplets in each block at the encoder input (each block contains $3N$ data bits). Two sets of the permutation parameters $\{p_0,\ p_1,\ p_2$ and $p_3\}$ are shown in Table 4.2. The permutation is as follows for $j = 0,\ ...,\ N-1$.

- If j mod. 4 = 0, then $p = 0$;
- If j mod. 4 = 1, then $p = N/2 + p_1$;
- If j mod. 4 = 2, then $p = p_2$;
- If j mod. 4 = 3, then $p = N/2 + p_3$;

$$i = (p_0 * j + p + 1)\ mod.\ N \tag{4.1}$$

The interleaving relations satisfy the odd/even rule (*i.e.*, when j is even, i is odd and vice-versa) that enables the puncturing patterns to be identical for both encoders.

Figure 4.3 shows the simulation results concerning the effect of different permutation parameters. It shows that using the parameter set P_1 results in better performance.

Frame size in triples	$P_1 = \{p_0, p_1, p_2, p_3\}$	$P_2 = \{p_0, p_1, p_2, p_3\}$
$N = 152$ (57 *bytes*)	{11,34,16,2}	{11,78,4,2}
$N = 224$ (84 *bytes*)	{23,114,8,118}	{19,108,4,116}
$N = 752$ (282 *bytes*)	{19,2,16,6}	{19,376,224,600}

Table 4.2. Triple-binary Code Permutation Parameters

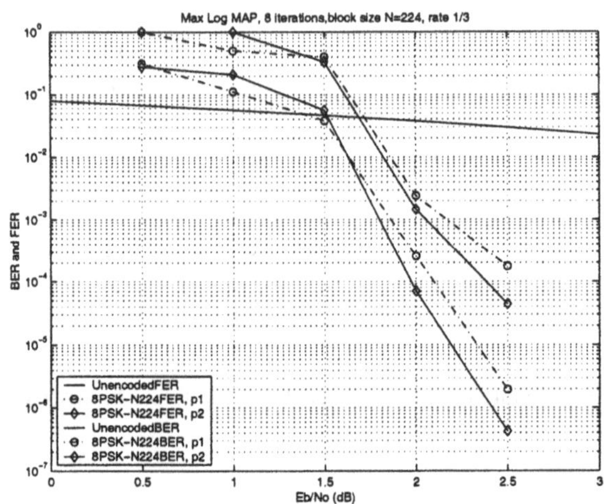

Figure 4.3. Performance of Frame Size N=224 (84 bytes) with Different Permutation Parameters

4.2.4 Puncturing Map, Order of Transmission and Mapping to 8PSK Constellation

Two code rates are defined for the triple-binary CRSC turbo code: $R = 1/3,\ and\ 2/3$. For rate $1/3$, the systematic bits and all encoded bits are transmitted, and rate $2/3$ is achieved through selectively deleting the parity bits (puncturing). The puncturing patterns of Table 4.3 are applied. This pattern is identical for both codes C_1 and C_2 (deletion is always done in triplets).

The order of transmission is in the natural order: all triplets (A, B, C) are transmitted first, followed by all triplets (W_1, Y_1, Z_1) that remain after puncturing and then all triplets (W_2, Y_2, Z_2) that remain after puncturing (see Figure 4.4).

Each triplet is mapped into one 8PSK constellation point as shown in Figure 4.5. In Figure 4.4, the columns with the systematic symbols and the columns with parity symbols are each mapped into one 8PSK constellation point, *i.e.*, systematic symbols (C, B, A), or parity symbols (Z_1, Y_1, W_1) and (Z_2, Y_2, W_2) correspond to (C_3, C_2, C_1). The signal shall be modulated using 8PSK, with baseband shaping. The output (C_3, C_2, C_1) of the channel encoder shall be

$$\text{Code rate 1/3} \quad \begin{matrix} Z \\ Y \\ W \end{matrix} \begin{bmatrix} 1 & 1 & 1 & 1 & 1 & 1 & 1 & 1 \\ 1 & 1 & 1 & 1 & 1 & 1 & 1 & 1 \\ 1 & 1 & 1 & 1 & 1 & 1 & 1 & 1 \end{bmatrix}$$

$$\text{Code rate 2/3} \quad \begin{matrix} Z \\ Y \\ W \end{matrix} \begin{bmatrix} 1 & 0 & 0 & 0 & 1 & 0 & 0 & 0 \\ 1 & 0 & 0 & 0 & 1 & 0 & 0 & 0 \\ 1 & 0 & 0 & 0 & 1 & 0 & 0 & 0 \end{bmatrix}$$

Table 4.3. Puncturing Patterns (Compared with Unpunctured Pattern) for Triple-binary CRSC Code. "1" = keep

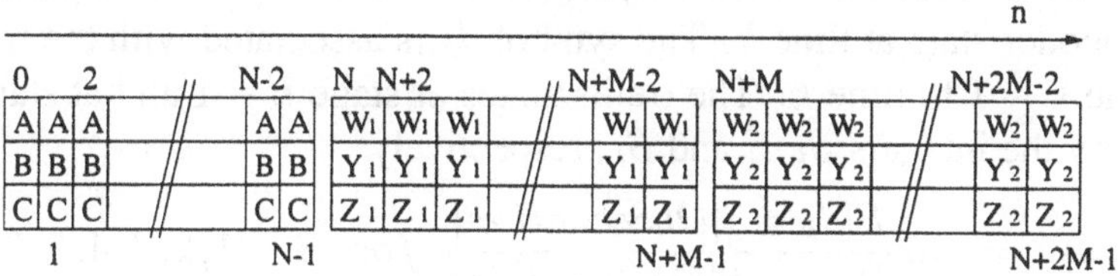

n

0 2 N-2 N N+2 N+M-2 N+M N+2M-2

A A A // A A | W_1 W_1 W_1 // W_1 W_1 | W_2 W_2 W_2 // W_2 W_2

B B B // B B | Y_1 Y_1 Y_1 // Y_1 Y_1 | Y_2 Y_2 Y_2 // Y_2 Y_2

C C C // C C | Z_1 Z_1 Z_1 // Z_1 Z_1 | Z_2 Z_2 Z_2 // Z_2 Z_2

1 N-1 N+M-1 N+2M-1

Figure 4.4. Encoded Blocks (Natural Order). M=N, Unpunctured; M=1/4N, Punctured.

mapped into the I channel and the Q channel as shown on the bottom of Figure 4.5.

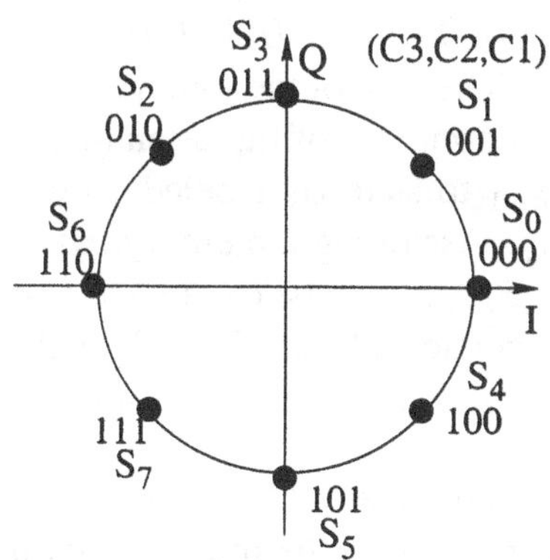

(C3, C2, C1) = ((C, Z1, or Z2), (B, Y1, or Y2), (A, W1, or W2)) = (I, Q)

$$\begin{array}{lll lll} S_0 = (000) & \rightarrow & (1,0) & S_4 = (100) & \rightarrow & (\frac{\sqrt{2}}{2}, \frac{\sqrt{2}}{2}) \\ S_1 = (001) & \rightarrow & (\frac{\sqrt{2}}{2}, \frac{\sqrt{2}}{2}) & S_5 = (101) & \rightarrow & (0,-1) \\ S_2 = (010) & \rightarrow & (-\frac{\sqrt{2}}{2}, \frac{\sqrt{2}}{2}) & S_6 = (110) & \rightarrow & (-1,0) \\ S_3 = (011) & \rightarrow & (0,1) & S_7 = (111) & \rightarrow & (-\frac{\sqrt{2}}{2}, -\frac{\sqrt{2}}{2}) \end{array}$$

Figure 4.5. Gray Mapping for 8PSK Constellation

4.3. Iterative Decoding Procedure

In DVB-RCS standard, each couple is mapped into one QPSK constellation point and every bit is mapped on the I channel and Q channel, respectively. Actually, the signal shall be sent without modification to the QPSK bit mapper (see Figure 3.7) and the iterative decoding performs with bit-by-bit input values. Due to the restrictions of the iterative decoding procedure, the interleaving and puncturing are done symbol-by-symbol for 8PSK constellation, so that, the decoding algorithm can be derived as a symbol-by-symbol Max-Log-MAP algorithm.

The trellis of the triple-binary feedback convolutional encoder that we use, has 16 states and each node has 8 symbol inputs and 8 symbol outputs. Let S_k be the encoder state at time k. The symbol d_k is associated with the transition from time $k-1$ to time k. The trellis states at stage $k-1$ and at stage k are indexed by the integers S_{k-1} and S_k, respectively.

$$\begin{aligned} L_i(d_k) &= \ln \frac{P[d_k = i|Observation]}{P[d_k = 0|Observation]} \quad for \quad i = 1,2,3,4,5,6,7 \\ &= \ln \frac{\sum_{d_k=i}^{(S_{k-1},S_k)} p(S_{k-1}, S_k, y_k)}{\sum_{d_k=0}^{(S_{k-1},S_k)} p(S_{k-1}, S_k, y_k)} \end{aligned} \tag{4.2}$$

The index pair (S_{k-1}, S_k) determines the information symbol d_k and the coded symbol x_k, where d_k and x_k are in GF(2^3) with elements {0, 1, 2, 3, 4, 5, 6, 7} that represent {000, 001, 010, 011, 100, 101, 110, 111}, respectively, from time $k-1$ to time k. The sum of the joint probabilities $p(S_{k-1}, S_k, y_k)$ in the numerator or in the denominator of Equation (4.2) is taken over all existing transitions from state S_{k-1} to state S_k labeled with the information symbols d_k and coded symbols x_k. Assuming a memoryless transmission channel, the joint probability $p(S_{k-1}, S_k, y_k)$ can be calculated using Equation (3.12). The forward and backward recursion of the MAP algorithm can also be written as Equations (3.13) and (3.14). The branch transition probabilities are also given by Equation (3.15).

In the same way, we denote the $\gamma_k(S_{k-1}, S_k)$ corresponding to the branch connecting a state S_{k-1} to S_k with the branch label $d_k = i$ as $\gamma_k^i(S_{k-1}, S_k)$, for $i = 0,1,2,3,4,5,6,7$ in the following equations. Then, the distribution of the received parity and systematic symbols are given by

$$\begin{aligned} & p[y_k|d_k = i] \\ = \; & p[y_k^s|x_k^s(i)] \cdot p[y_k^p|x_k^p(i, S_{k-1}, S_k)] \\ = \; & \frac{1}{\pi \cdot N_0} exp\{-\frac{E_s}{N_0}[(y_k^{s,I} - x_k^{s,I}(i))^2 + (y_k^{s,Q} - x_k^{s,Q}(i))^2]\} \\ & \cdot \frac{1}{\pi \cdot N_0} exp\{-\frac{E_s}{N_0}[(y_k^{p,I} - x_k^{p,I}(i, S_{k-1}, S_k))^2 + (y_k^{p,Q} - x_k^{p,Q}(i, S_{k-1}, S_k))^2]\} \\ = \; & B_k \cdot exp\{\frac{1}{2} L_c \cdot [y_k^{s,I} \cdot x_k^{s,I}(i) + y_k^{s,Q} \cdot x_k^{s,Q}(i)]\} \end{aligned}$$

$$\cdot exp\{\frac{1}{2}L_c \cdot [y_k^{p,I} \cdot x_k^{p,I}(i, S_{k-1}, S_k) + y_k^{p,Q} \cdot x_k^{p,Q}(i, S_{k-1}, S_k)]\} \tag{4.3}$$

where y_k^s, y_k^p represent the received systematic and parity symbols, and $y_k^{s,I}$, $y_k^{s,Q}$, $y_k^{p,I}$, $y_k^{p,Q}$ represent the received symbol values that are transmitted through the I and Q channels, respectively; $x_k^s(i)$, $x_k^p(i, S_{k-1}, S_k)$ represent the systematic and parity symbols for $i = 0, 1, 2, 3, 4, 5, 6, 7$, and $x_k^{s,I}(i)$, $x_k^{s,Q}(i)$, $x_k^{p,I}(i, S_{k-1}, S_k)$, $x_k^{p,Q}(i, S_{k-1}, S_k)$ represent the symbols of codeword mapped to 8PSK constellation, respectively.

Here,

$$\begin{aligned} B_k &= (\frac{1}{\pi \cdot N_0})^2 exp\{-\frac{E_s}{N_0}[(y_k^{s,I})^2 + (x_k^{s,I}(i))^2 + (y_k^{s,Q})^2 + (x_k^{s,Q}(i))^2 \\ &\quad +(y_k^{p,I})^2 + (x_k^{p,I}(i, S_{k-1}, S_k))^2 + (y_k^{p,Q})^2 + (x_k^{p,Q}(i, S_{k-1}, S_k))^2]\} \end{aligned} \tag{4.4}$$

4.3.1 Max-Log-MAP Algorithm for Triple-binary Codes

Symbol-by-symbol Max-Log-MAP algorithm is derived for the triple-binary codes with higher order modulation 8PSK using Gray mapping. First, find the logarithm of the branch metrics as

$$\begin{aligned} &\overline{\gamma_k^i}(S_{k-1}, S_k) \\ &= \ln \gamma_k^i(S_{k-1}, S_k) \\ &= \ln P(y_k|d_k) \cdot P(d_k) \\ &= \frac{1}{2}L_c \cdot [y_k^{s,I} \cdot x_k^{s,I}(i) + y_k^{s,Q} \cdot x_k^{s,Q}(i)] + \ln P(d_k) + K \\ &\quad +\frac{1}{2}L_c \cdot [y_k^{p,I} \cdot x_k^{p,I}(i, S_{k-1}, S_k) + y_k^{p,Q} \cdot x_k^{p,Q}(i, S_{k-1}, S_k)] \end{aligned} \tag{4.5}$$

where constant K includes the constant and common terms that are cancelled in comparisons at later stages. Next, compute $\overline{\alpha_k}(S_k)$ and $\overline{\beta_{k-1}}(S_k)$ as

$$\begin{aligned} \overline{\alpha_k}(S_k) &= \ln \sum_{S_{k-1}} e^{\overline{\gamma_k}(S_{k-1},S_k)} \cdot e^{\overline{\alpha_{k-1}}(S_{k-1})} \\ &\approx \max_{S_{k-1}}[\overline{\gamma_k}(S_{k-1}, S_k) + \overline{\alpha_{k-1}}(S_{k-1}) \end{aligned} \tag{4.6}$$

Similarly

$$\begin{aligned} \overline{\beta_{k-1}}(S_{k-1}) &= \ln \sum_{S_k} e^{\overline{\gamma_k}(S_{k-1},S_k)} \cdot e^{\overline{\beta_k}(S_k)} \\ &\approx \max_{S_k}\{[\overline{\gamma_k}(S_{k-1}, S_k) + \overline{\beta_k}(S_k)] \end{aligned} \tag{4.7}$$

For the iterative decoding of the circular trellis, we have

$$\begin{aligned} \overline{\alpha_0}(S_0) &= \overline{\alpha_N}(S_N) \quad for\ \forall S_0 \\ \overline{\beta_N}(S_N) &= \overline{\beta_0}(S_0) \quad for\ \forall s_N \end{aligned} \tag{4.8}$$

The log-likelihood ratios are expressed as,

$$\begin{aligned} L_i(d_k) &= \ln \frac{\sum_{d_k=i}^{(S_{k-1},S_k)} \gamma_k^i(S_{k-1},S_k)\cdot\alpha_{k-1}(S_{k-1})\cdot\beta_k(S_k)}{\sum_{d_k=0}^{(S_{k-1},S_k)} \gamma_k^0(S_{k-1},S_k)\cdot\alpha_{k-1}(S_{k-1})\cdot\beta_k(S_k)} \\ &= \ln \frac{\sum_{d_k=i}^{(S_{k-1},S_k)} e^{\overline{\gamma_k^i}(S_{k-1},S_k)}\cdot e^{\overline{\alpha_{k-1}}(S_{k-1})}\cdot e^{\overline{\beta_k}(S_k)}}{\sum_{d_k=0}^{(S_{k-1},S_k)} e^{\overline{\gamma_k^0}(S_{k-1},S_k)}\cdot e^{\overline{\alpha_{k-1}}(S_{k-1})}\cdot e^{\overline{\beta_k}(S_k)}} \\ & \qquad\qquad for\ \ i=1,2,3,4,5,6,7 \end{aligned} \tag{4.9}$$

Therefore, to compute the log-likelihood ratios, we follow the equation (4.9) and take max-function as

$$\begin{aligned} L_i(\hat{d}_k) &\approx \max_{(s_{k-1},s_k)} [\overline{\gamma_k^i}(S_{k-1},S_k)+\overline{\alpha_{k-1}}(S_{k-1})+\overline{\beta_k}(S_k)] \\ &\quad - \max_{(s_{k-1},s_k)} [\overline{\gamma_k^0}(S_{k-1},S_k)+\overline{\alpha_{k-1}}(S_{k-1})+\overline{\beta_k}(S_k)] \\ & \qquad\qquad for\ i=1,2,3,4,5,6,7 \end{aligned} \tag{4.10}$$

Moreover, to separate the Log-likelihood ratios into intrinsic, systematic and extrinsic information, define:

$$\begin{aligned} \gamma_k^{i(e)}(S_{k-1},S_k) &= exp\{\frac{1}{2}L_c\cdot[y_k^{p,I}\cdot x_k^{p,I}(i,S_{k-1},S_k) \\ &\quad +y_k^{p,Q}\cdot x_k^{p,Q}(i,S_{k-1},S_k)]\}\cdot A_k \end{aligned} \tag{4.11}$$

where

$$\begin{aligned} A_k &= \frac{1}{\pi\cdot N_0} exp\{-\frac{E_s}{N_0}[(y_k^{p,I})^2+(x_k^{p,I}(i,S_{k-1},S_k))^2 \\ &\quad +(y_k^{p,Q})^2+(x_k^{p,Q}(i,S_{k-1},S_k))^2]\} \end{aligned} \tag{4.12}$$

So, the logarithm of the branch transition operation reduces to the expression with

$$\begin{aligned} & \overline{\gamma_k^{i(e)}(S_{k-1},S_k)} \\ =\ & \ln\gamma_k^{i(e)}(S_{k-1},S_k) \\ =\ & \tfrac{1}{2}L_c\cdot[y_k^{p,I}\cdot x_k^{p,I}(i,S_{k-1},S_k)+y_k^{p,I}\cdot x_k^{p,Q}(i,S_{k-1},S_k)]+K' \end{aligned} \tag{4.13}$$

where constant K' includes the constant and common terms that are cancelled in comparisons at later stages.

In another way, find the Log-likelihood ratios as

$$\begin{aligned}
L_i(\hat{d}_k) &= \ln \frac{\sum_{d_k=i}^{(S_{k-1},S_k)} \gamma_k^i(S_{k-1},S_k)\cdot\alpha_{k-1}(S_{k-1})\cdot\beta_k(S_k)}{\sum_{d_k=0}^{(S_{k-1},S_k)} \gamma_k^0(S_{k-1},S_k)\cdot\alpha_{k-1}(S_{k-1})\cdot\beta_k(S_k)} \quad \textit{for } i=1,2,3,4,5,6,7 \\
&= \ln \frac{exp\{\frac{1}{2}L_c\cdot[y_k^{s,I}\cdot x_k^{s,I}(i)+y_k^{s,Q}\cdot x_k^{s,Q}(i)]\}\cdot P[d_k=i]}{exp\{\frac{1}{2}L_c\cdot[y_k^{s,I}\cdot x_k^{s,I}(0)+y_k^{s,Q}\cdot x_k^{s,Q}(0)]\}\cdot P[d_k=0]} \\
&\quad + \ln \frac{\sum_{d_k=i}^{(S_{k-1},S_k)} \gamma_k^{i(e)}(S_{k-1},S_k)\cdot\alpha_{k-1}(S_{k-1})\cdot\beta_k(S_k)}{\sum_{d_k=0}^{(S_{k-1},S_k)} \gamma_k^{0(e)}(S_{k-1},S_k)\cdot\alpha_{k-1}(S_{k-1})\cdot\beta_k(S_k)} \\
&= \{\frac{1}{2}L_c\cdot[y_k^{s,I}\cdot x_k^{s,I}(i)+y_k^{s,Q}\cdot x_k^{s,Q}(i)] - \frac{1}{2}L_c\cdot[y_k^{s,I}\cdot x_k^{s,I}(0)+y_k^{s,Q}\cdot x_k^{s,Q}(0)]\} \\
&\quad + \underbrace{\ln \frac{P[d_k=i]}{P[d_k=0]}}_{L_i(d_k)} + \underbrace{\ln \frac{\sum_{d_k=i}^{(S_{k-1},S_k)} \gamma_k^{i(e)}(S_{k-1},S_k)\cdot\alpha_{k-1}(S_{k-1})\cdot\beta_k(S_k)}{\sum_{d_k=0}^{(S_{k-1},S_k)} \gamma_k^{0(e)}(S_{k-1},S_k)\cdot\alpha_{k-1}(S_{k-1})\cdot\beta_k(S_k)}}_{L_i^e(\hat{d}_k)}
\end{aligned} \tag{4.14}$$

So, the extrinsic information is

$$\begin{aligned}
L_i^e(\hat{d}_k) &= L_i(\hat{d}_k) - \frac{1}{2}L_c\cdot[y_k^{s,I}\cdot x_k^{s,I}(i)+y_k^{s,Q}\cdot x_k^{s,Q}(i)] \\
&\quad + \frac{1}{2}L_c\cdot[y_k^{s,I}\cdot x_k^{s,I}(0)+y_k^{s,Q}\cdot x_k^{s,Q}(0)] - \ln\frac{P[d_k=i]}{P[d_k=0]}
\end{aligned} \tag{4.15}$$

The computation of the symbol probabilities for the next decoder is as follows for $i = 1, 2, 3, 4, 5, 6, 7$ from previous decoder,

$$L_i(d_k) = L_i^e(\hat{d}_k) = \ln\frac{P[d_k=i]}{P[d_k=0]} \tag{4.16}$$

Since

$$\begin{cases} L_0^e(\hat{d}_k) = \ln\frac{P[d_k=0]}{P[d_k=0]} = \ln 1 = 0 \\ L_i^e(\hat{d}_k) = \ln\frac{P[d_k=i]}{P[d_k=0]} \end{cases} \tag{4.17}$$

then

$$P[d_k=i] = e^{L_i^e(\hat{d}_k)}\cdot P[d_k=0] \tag{4.18}$$

and

$$\sum_{i=0}^{7} P[d_k=i] = 1 \tag{4.19}$$

We have,

$$\begin{cases} P[d_k=0] = \frac{1}{1+\sum_{i=1}^{7} e^{L_i^e(\hat{d}_k)}} \\ P[d_k=i] = \frac{e^{L_i^e(\hat{d}_k)}}{1+\sum_{i=1}^{7} e^{L_i^e(\hat{d}_k)}} \end{cases} \tag{4.20}$$

Using max-function

$$\begin{cases} \ln P[d_k = 0] &= -\max[0, L_i^e(\hat{d}_k)] \\ \ln P[d_k = i] &= L_i^e(\hat{d}_k) - \ln P[d_k = 0] \end{cases} \quad for\ i = 1, 2, 3, 4, 5, 6, 7 \tag{4.21}$$

4.3.2 Initialization and the Final Decision

We have no *a priori* information available for the first iteration, and we initialize

$$L_i(d_k = i) = 0 \quad for\ i = 0, 1, 2, 3, 4, 5, 6, 7 \tag{4.22}$$

according to Equation (4.20), we have

$$\begin{cases} P[d_k = 0] &= \frac{1}{1+\sum_{i=1}^{7} e^{L_i^e(\hat{d}_k)}} = \frac{1}{8} \\ \\ P[d_k = i] &= \frac{e^{L_i^e(\hat{d}_k)}}{1+\sum_{i=1}^{7} e^{L_i^e(\hat{d}_k)}} = \frac{1}{8} \end{cases} \tag{4.23}$$

Using max-function:

$$\begin{cases} \ln P[d_k = 0] &= -\max[0, L_i^e(\hat{d}_k)] = 0 \\ \ln P[d_k = i] &= L_i^e(\hat{d}_k) - \ln P[d_k = 0] = 0 \end{cases} \quad for\ i = 1, 2, 3, 4, 5, 6, 7 \tag{4.24}$$

Similarly, because of equal-likelihood assumption for all symbols, we have

$$\begin{array}{rcl} \alpha_0(s_0) &=& 1 \quad for\ \forall s_0 \\ \beta_N(s_N) &=& 1 \quad for\ \forall s_N \end{array} \tag{4.25}$$

take logarithm

$$\begin{array}{rcccl} \overline{\alpha_0}(S_0) &=& \ln \alpha_0(s_0) &=& 0 \quad for\ \forall s_0 \\ \overline{\beta_N}(S_N) &=& \ln \beta_N(s_N) &=& 0 \quad for\ \forall s_N \end{array} \tag{4.26}$$

and initialize $\overline{\gamma_0^i} = 0$.

The reliability value of the channel

$$L_c = 4a \cdot \frac{E_s}{N_0} = 4a \cdot R_c \cdot \frac{E_b}{N_0} \tag{4.27}$$

where R_c is the code rate.

After several decoding iterations, the decisions are made according to:

$$\begin{cases} \hat{d}_k = i & if\ L(\hat{d}_k) = L_i(\hat{d}_k)\ and\ L_i(\hat{d}_k) > 0 \\ & for\ i = 1, 2, 3, 4, 5, 6, 7 \\ \hat{d}_k = 0 & else \end{cases} \tag{4.28}$$

where

$$L(\hat{d}_k) = \max(L_i(\hat{d}_k)) \quad for\ i = 1, 2, 3, 4, 5, 6, 7 \tag{4.29}$$

4.4. Simulation Results

Three different frame sizes are investigated and Figure 4.6 shows the performance. The curves on the left of Figure 4.6 correspond to the code rate of $1/3$, *i.e.*, a bandwidth efficiency of 1bit/s/Hz. The curves on the right correspond to the code rate of $2/3$, *i.e.*, a bandwidth efficiency of 2bits/s/Hz.

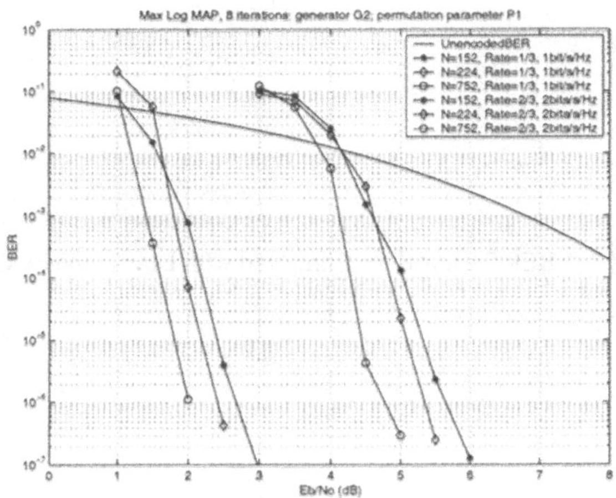

Figure 4.6. Performance of Three Different Frame Sizes with Different Bandwidth Efficiency

Figure 4.7 compares the BER performance of triple-binary CRSC codes with double-binary CRSC codes in DVB-RCS standard. The curves with solid line on the left of Figure 4.7 correspond to the double-binary/QPSK codes, a code rate of $1/2$, block size $N = 228$ (57 bytes) using **level 1** and **level 2** interleaving. The dashed line is for triple-binary/8PSK code, a code rate of $1/3$ and one **level 2** uniform interleaving. All of these curves are for a bandwidth efficiency of 1bit/s/Hz. The performance is very close.

The curve with solid line on the right of Figure 4.7 is the ATM cell (53 bytes) with a code rate of $6/7$ and a bandwidth efficiency of 1.7 bits/s/Hz. The dashed line with star is the punctured triple-binary CRSC code with a code rate $2/3$ and a bandwidth efficiency of 2bits/s/Hz, $N = 152$ (57 bytes). The performance of triple-binary/8PSK code is better than double-binary/QPSK code at higher signal-to-noise ratios with higher bandwidth efficiency. For an information theoretical explanation of this result, see Figure 1.5 and the corresponding discussion in Chapter 1.

Figure 4.8 compares the two coding schemes in terms of Frame Error Rate (FER).

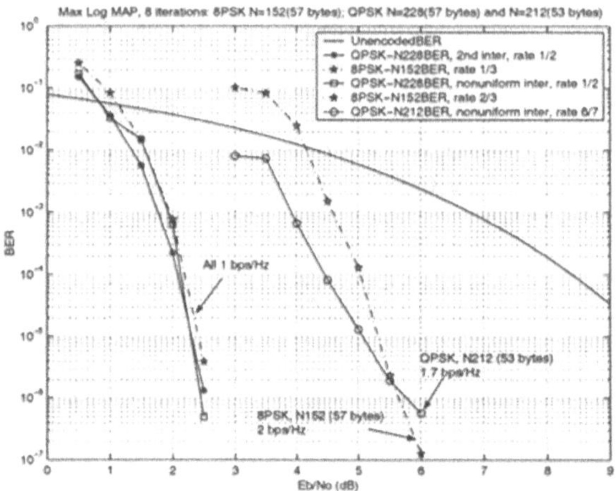

Figure 4.7. BER Performance Compared with Double-binary CRSC Codes

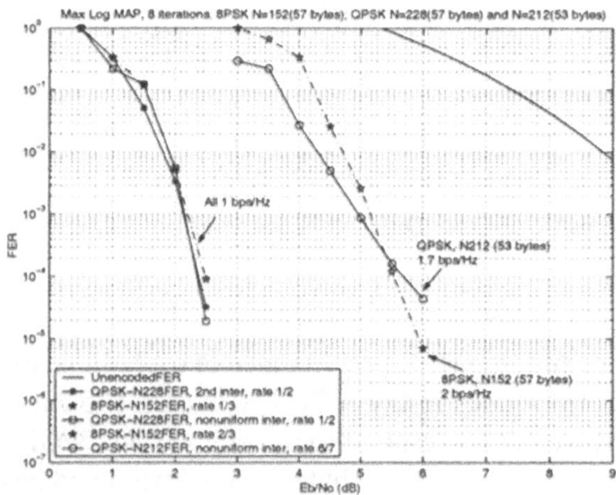

Figure 4.8. FER Performance Compared with Double-binary CRSC Codes

4.5. Turbo Trellis Coded Modulation Schemes

Trellis coded modulation (TCM), introduced by Ungerboeck [101], [102] and [103] is a very effective method for reducing the required power without any increase in the bandwidth requirement. The innovative aspect of TCM is the concept that encoding and modulation should not be treated as separate entities, but rather, as a unique operation. TCM schemes have been applied to telephone, satellite and microwave digital radio channels, where coding gains of the order of 3-6dB are obtained with no loss of bandwidth or data rate [60].

Turbo codes can achieve remarkable error performance at a low signal-to-noise ratio close to the Shannon capacity limit. However, the powerful binary coding schemes are not suitable for bandwidth limited communication systems. In order to achieve simultaneously large coding gains and high bandwidth efficiency, a general method is to combine turbo codes with trellis coded modulation. *Turbo Trellis Coded Modulation* (TTCM) proposed in [104] is the extension of turbo codes where the component codes are replaced by Ungerboeck TCM codes in the recursive systematic forms to retain the advantages of both classical turbo codes and TCM codes. The TTCM scheme, is called *Parallel Concatenated Trellis Coded-Modulation* (PCTCM) or *Serial Concatenated Trellis Coded-Modulation* (SCTCM) according to its encoder structure. With the remarkable performance of turbo codes, it is natural to combine turbo codes with multilevel modulation schemes in order to obtain large coding gains and high bandwidth efficiency over both AWGN and fading channels [105]. In this section, we discuss several bandwidth efficient turbo coding schemes.

4.5.1 Pragmatic Binary Turbo Coded Modulation

The first attempt in combining turbo codes with multilevel modulation was described in [109] and is called "pragmatic" approach to TCM. In this approach a Gray mapper is used after binary turbo encoder for multilevel modulation. The coding and modulation are separated processes and hence it is actually not a coded-modulation scheme. Decoding relies on the binary turbo decoder, hence the term "pragmatic". Figure 4.9 shows the association of a binary turbo code

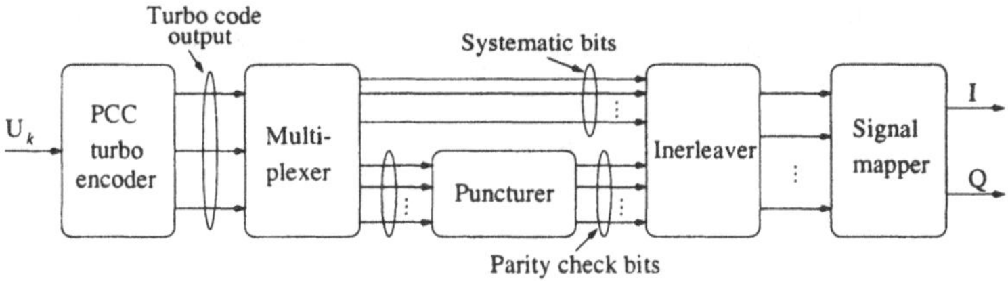

Figure 4.9. Association of Turbo Codes with Multilevel Modulations

with M-level modulation (MPSK, M-QAM). The standard turbo code uses two rate 1/2 RSC codes as constituent codes.

The parity check bits at the output of the constituent codes are denoted as C^1 and C^2, respectively. The puncturing function is inserted at the output of the standard turbo code and thus it is possible to obtain a large code family with various rates $R = (m - \tilde{m})/m$, where $\tilde{m}$ is the number of parity bits and m is the total number of bits that are Gray mapped into a complex signal symbol to be transmitted over the channel.

In the approach of [56], binary RSC component codes in binary turbo code are replaced by Ungerboeck TCM codes to retain the advantages of both classical turbo code and TCM code. At the receiver (see Figure 4.10), the log-likelihood value is calculate for every encoded binary digit corrupted by the channel noise. Then, the sequence of the bit log-likelihood values is deinterleaved and demultiplexed before being passed to the turbo decoder based on MAP, Log-MAP, Max-Log-MAP algorithm, or SOVA.

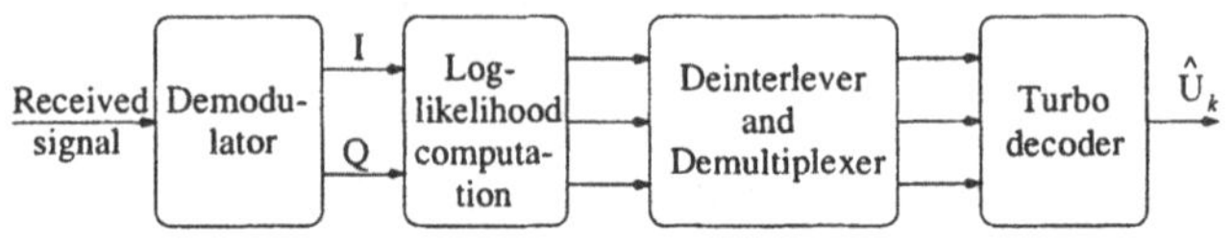

Figure 4.10. Decoder for Concatenated PCCC/TCM Code

4.5.2 Turbo Trellis Coded Modulation

Motivated by the multilevel coding scheme of [107], a method of combining turbo codes with multilevel modulation scheme was introduced in [56] and [108]. It involved the parallel concatenation of two recursive Ungerboeck type trellis codes with M-ary signal constellation. Figure 4.11 shows the encoder structure comprising of two recursive convolutional encoders each followed by a signal mapper and linked by a symbol interleaver. The switch at the output of the turbo encoder punctures the code by selecting the odd symbols of one mapper and the even symbols of the other.

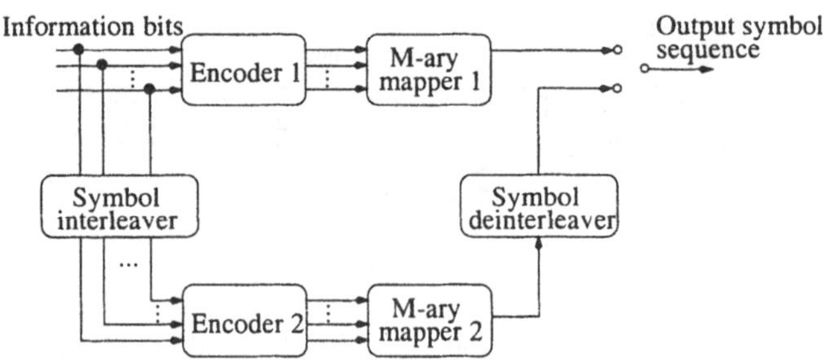

Figure 4.11. TTCM Encoder

The decoder structure is shown in Figure 4.12. We showed in Equation 2.23 that, for a conventional turbo code, the output of each constituent decoder can be written as the sum of three terms, i.e., the *systematic*, the *a priori* and the *extrinsic* components and only the extrinsic part is passed to the other decoder. It is seen from Figure 4.11 that, for this scheme, both the information and parity bits are mapped to the same modulation symbol and subsequently are affected by the same noise sample. As a result, the systematic part cannot be separated from the extrinsic part. However, it is possible to split the output of the decoder into two parts: 1) an *a priori* part and 2) a part corresponding to the *extrinsic* and *systematic* information. Only the second part is passed to the other decoder to be used as the *a priori* information. This part is obtained by subtracting the *a priori* information (the soft input of the decoder) from its soft output. The initial *a priori* probabilities of all information symbols are assumed to be equal at the beginning of the decoding. Each decoder ignores the symbols that are not from its own corresponding encoder. For example, if the puncturer picks the odd numbered outputs of the first mapper and the even numbered outputs of the second then, the first decoder ignores the even numbered noisy inputs and the second decoder ignores the odd ones. The only input in these instances is the information received from the other decoder. For more detail about the decoding procedure including the initialization, the reader is referred to [108].

Using TTCM, a coding gain of about 1.7 dB can be achieved compared to a conventional TCM, at an error rate in the vicinity of 10^{-4}. This means that turbo TCM achieves a performance close to the Shannon capacity on an AWGN channel [44].

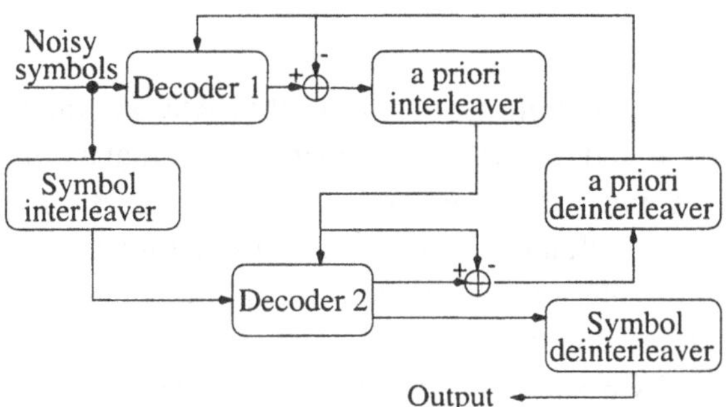

Figure 4.12. TTCM Decoder

In [108], two Ungerboeck-type 8PSK mappers in combination with two recursive systematic component codes with memory $v = 3$ are employed. The MAP algorithm is applied and block size is $N = 1024$ (256 bytes, 2048 bits) for the code rate $1/3$ in this TTCM scheme. The bandwidth efficiency is 2 bit/s/Hz at 5.9 dB (E_s/N_0). Figure 4.13 shows the comparison of the triple-binary/8PSK code with this TTCM code.

The triple-binary/8PSK code with frame size, $N = 752$ (282 bytes, 2256 bits), a code rate of $2/3$, the memory $v = 4$, uses Max-Log-MAP algorithm. This simulation result shows that the performance of the triple-binary/8PSK and TTCM are very close. Moreover, at receiver, the triple-binary/8PSK coding scheme avoids the calculation of the log-likelihood value for every encoded binary digit and the Max-Log-MAP algorithm is simple and easy to be implemented.

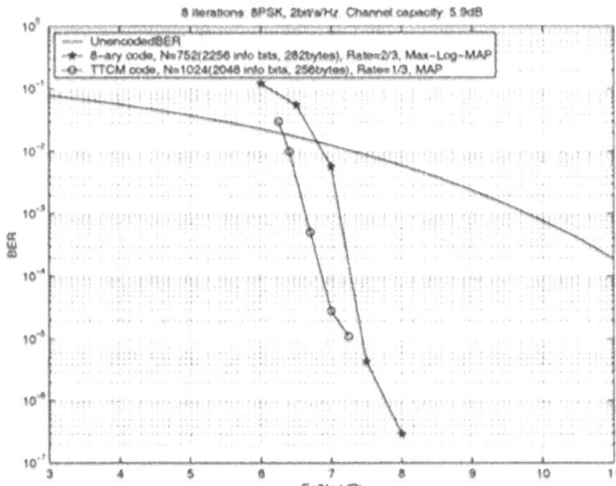

Figure 4.13. Triple-binary CRSC Code Compared with TTCM. Both for 8PSK Modulation and Bandwidth Efficiency: 2bps/Hz at 5.9 dB (E_s/N_0)

4.6. Summary

Circular Recursive Systematic Convolutional (CRSC) component codes, non-uniform permutation and different puncturing maps make double-binary turbo codes efficient and provide better error-correcting performance than binary codes for equivalent implementation complexity. Motivated by the advantages of double-binary convolutional turbo codes, triple-binary codes are designed to be used with 8PSK modulation.

While the DVB-RCS standard has an excellent performance, its bandwidth efficiency is limited by the puncturing and QPSK modulation to less than 2 bits/s/Hz. In order to achieve bandwidth efficiencies of 2 bits/s/Hz and higher, one needs coding and modulation schemes with M-ary alphabet (M > 4). A new triple-binary CRSC code is designed for 8PSK modulation in order to increase the bandwidth efficiency. This triple-binary code still has the features of the CRSC codes, which avoid the degradation of the spectral efficiency. Two circulation states have to be determined and the sequence has to be encoded four times. The different permutations (interleavings) can be obtained using generic equations with only a restricted number of parameters.

Triple-binary CRSC codes inherit most of the advantages of double-binary CRSC codes, however, they are more flexible and efficient for encoding blocks of data. Especially, the 8-ary alphabet of triple-binary turbo codes when combined with 8PSK achieve higher code rate and more than 2 bits/s/Hz bandwidth efficiency. The bitwise interleaver known from classical binary turbo codes is replaced by an interleaver operating on a group of bits and the equations of permutation parameters are chosen to be the same as the **level 2** interleaver of DVB-RCS code. The structure of the iterative decoder and the symbol-by-symbol Max-Log-MAP algorithm are derived for non-binary trellises to avoid numerical problems and reduce the decoding complexity due to a set of constraints on the component code, interleaver and 8PSK symbol mapping. This preliminary study of 8-ary triple-binary CRSC codes show their potential as an alternative for more bandwidth efficient transmission of data via satellites without an increase in the required bandwidth. Moreover, turbo trellis coded modulation was discussed. Simulation results show that the performance of the triple-binary CRSC codes are close to that of TTCM scheme.

Chapter 5

BLOCK TURBO CODES

5.1. Introduction

A *Block Turbo Code* (BTC) is a concatenated block code decoded with iterative decoding technique. There can be both serial and parallel concatenated codes. In some literatures, a serial concatenated code based on the idea of 2-dimensional product code is called a *Turbo Product Code* (TPC). There are two different *Soft-Input Soft-Output* (SISO) decoding methods for BTC, viz., the *Trellis*-based algorithm [7] and the *Algebraic* decoding based algorithm or *Augmented List Decoding* algorithm [128],[126] and [123].

We divide our literature review on the topic of BTCs into two parts based on the decoding method used. After a brief review of the BTCs, serial and parallel concatenated block codes with block interleaver are introduced. Then, serial and parallel iterative decoding ideas are discussed and, two SISO decoding algorithms are presented. One is based on the algebraic algorithm, i.e. modified Chase-II algorithm. The other is the trellis-based algorithm.

5.1.1 Trellis-Based Decoding

In trellis-based iterative algorithm, an extension of Bahl-Cocke-Jelinek-Raviv (BCJR) algorithm [57], the state sequence of a discrete-time finite state Markov process in a memoryless channel is estimated. Such a process can be represented by a trellis diagram. Both convolutional and block codes have a trellis diagram representation. Lodge et al. [4] presented the separable MAP-filters approach for decoding the multi-dimensional product codes and extended it to concatenated convolutional codes with interleaver. The extrinsic information, called the refinement factor, is passed from one decoding process to another through an iterative process. This paper was published one year before the introduction of Turbo Code by Berrou et al. [6]. It was also presented at the same conference [5] that Berrou et al. presented their famous paper. This means that BTC was developed even before TCC (*Turbo Convolutional Code*). In [119], the MAP decoding algorithm in log domain was developed. Then, Hagenauer et al. [7] presented a clear concept and a solid mathematical framework for iterative decoding of both convolutional and linear block codes using MAP algorithm and its variants. Moreover, the stopping criterion using cross-entropy was investigated for the sake of reducing complexity. The use of BTCs in a concatenated scheme was presented by Y. Liu and S. Lin in [120] using Reed-Solomon code as the outer code and Hamming turbo product code as the inner code. In [120], a new stopping criterion is proposed and applied to inner iterative decoding and the effect of parallel and serial decoding of turbo decoder is investigated and it is shown that the parallel decoding outperforms the serial decoding. In addition, in [121] Y. Liu et al. propose the trellis-based MAP decoding algorithm based on the sectionalized trellis of linear block codes. An optimal sectionalized trellis is considered as the best trellis in the sense of minimizing the number of multiplication operations. In [121], the optimal sectionalization of Reed-Muller (RM) codes are found. The analysis of the computational complexity and storage space are investigated. Parallel MAP decoding algorithm is also considered for RM codes by decomposing the trellis structure into identically parallel sub-trellises without cross connections among them and without exceeding the maximum state complexity of the trellis. By doing this , the decoding delay is reduced and the decoding process is speeded up making it suitable for hardware implementation.

It is important here to note the difference between this serial concatenated scheme and the more conventional scheme presented in Chapter 2. This scheme, except for the parity on parity bits, has no other difference with the parallel concatenation to be discussed next.

5.1.2 Augmented List Decoding

In this turbo decoding category, a list of candidate codewords are produced with different methods such as the Chase-II algorithm [122] used in [128], the Pseudo-Maximum-Likelihood (PLM) algorithm used in [123] and the Fang-Battail-Buda-Algorithm (FBBA) used in [126]. List decoding is considered as soft decision decoding of linear block codes because the list of candidate codewords is obtained from the channel information (soft information).

Pyndiah and his co-authors published a series of papers [128],[129],[10],[130],[131]. They used a product code or a serial concatenation of block codes with a block interleaver. The decoding algorithm used in SISO decoders is based on a modified Chase algorithm. The main idea of this algorithm is to reduce the number of the reviewed codewords in a set of highly probable codewords by using channel information. First, the set of error patterns, E is produced based on the reliability of the received sequence, then the set of test patterns, T is generated, where $\vec{t} = \vec{e} + \vec{h}$, $\vec{t} \in T$, $\vec{e} \in E$ and $\vec{h}$ is the hard decision vector of a received sequence. Each test pattern is decoded using an algebraic decoder and the set of candidate codewords is generated. The decision codeword is the codeword that possesses the highest correlation with the received sequence among the candidate code words. The soft-output of a given bit is calculated from the received vector, the decision codeword and the competing codeword or the codeword in the candidate set having the higher correlation and whose bit at a given position is different from the bit at that position of decision codeword.

The performance of BTCs using BCH codes as their component codes over a Gaussian channel was presented in [128]. The results show the attractiveness of BTCs for the applications that require very good performance with high code rates $R > 0.8$. The extension of this paper is presented in [10] with results for both AWGN and Rayleigh fading channels. It is shown that more than 98% of channel capacity can be achieved with a high code rate. A further investigation using Reed-Solomon codes as component codes was presented in [130] with an attempt to apply BTC to data storage applications.

The most significant drawback of turbo decoder is its complexity; thus, [129] presents the methods of reducing complexity of turbo product codes by reducing the number of test patterns and using the previous decision codeword as the competing codeword for the next iteration. Results show that the complexity is reduced almost by a factor of ten compared to [128] with a performance degradation of 0.7 dB. In [132], fast Chase algorithm is proposed by ordering test patterns before feeding them to the algebraic decoder in such a way that the operations in syndrome and metric calculation are reduced without performance degradation. Some recent improvements on the BTC in performance and implementation matters are presented in [131], [133] and [134]. The application of block turbo codes in wireless packet transmission is presented in [135], [126]. In [135], the PLM algorithm is used, whereas in [126] the FBBA algorithm is applied and *Unequal Error Protection* (UEP) property of *Generalized Turbo Product Code* (GTPC) is also introduced. Some details about the applications of BTCs used in satellite and wireless communications are given in Chapter 7.

5.2. Concatenated Block Codes with Block Interleaver

In this book, only the concatenation of block codes with *2-Dimensional* (2D) array information is considered. The well-known Product code [34] is an example of a serial concatenated block codes.

5.2.1 Serial Concatenated Block Codes

Figure 5.1 shows a serial concatenated encoder consisting of systematic outer and inner encoders. The information data is encoded by an outer encoder where the output is a coded sequence. Then the sequence is scrambled by an interleaver and fed to an inner encoder. In this book, we consider the specific case of serial concatenated block codes with block interleaver known as "Prod-

Parameters	C_1	C_2	$C_1 \times C_2$
Code Length	n_1	n_2	$n_1 \times n_2$
Information Length	k_1	k_2	$k_1 \times k_2$
Minimum Distance	d_1	d_2	$d_1 \times d_2$
Code Rate	R_1	R_2	$R_1 \times R_2$

Table 5.1. Parameters of a Product Code

Figure 5.1. The Serial Concatenated Block Codes

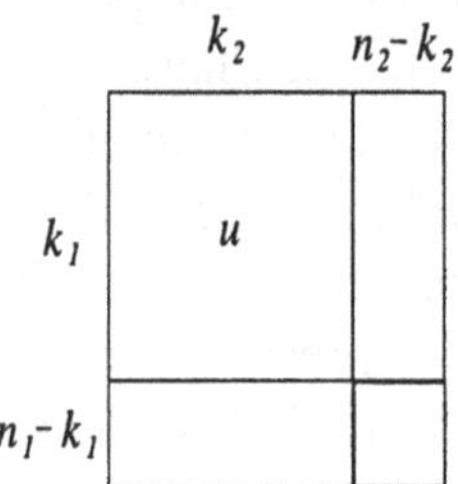

Figure 5.2. Product Code

uct Code". In a block interleaver, data is written row wise from left to right and from top to bottom and read out column wise from top to bottom and from left to right. The following is the principle of a product code.

A product code is a multidimensional block code, it may be 2-Dimensional (2D) or *3-Dimensional* (3D) or have a higher dimension. It is based on the concept of constructing a long block code with moderate decoding complexity by combining shorter codes.

Let's consider a 2D code consisting of a linear (n_1, k_1, δ_1) block code C_1 with rate R_1 and another linear (n_2, k_2, δ_2) block code C_2 with rate R_2. Here n_i, k_i, δ_i, $i = 1, 2$ are code length, information length and minimum distance. We use the notation $C_1 \times C_2$ to represent the product code constructed from component codes C_1 and C_2. The parameters of the product code (n_p, k_p, δ_p) are the product of parameters of the elementary codes and they are presented in Table 5.1.

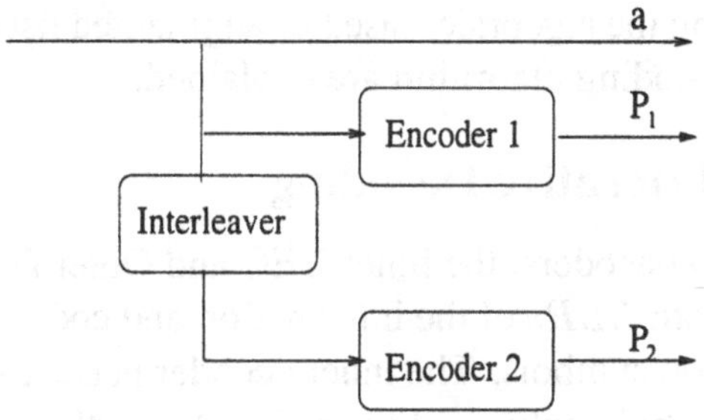

Figure 5.3. The Parallel Concatenated Block Code

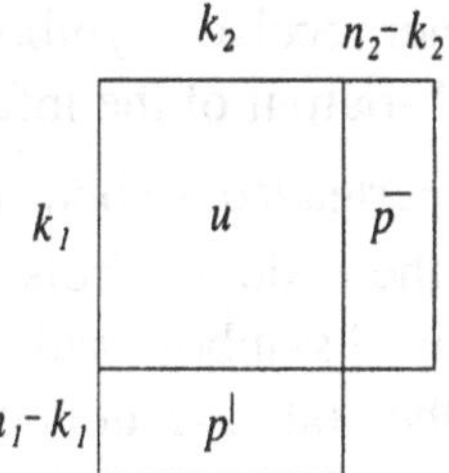

Figure 5.4. The Parallel Concatenated Block Code

Figure 5.2 shows the product code and its parameters. It is noted that the resulting code obtains larger minimal distance than the distance of its component codes. The decoding of a product code consists in decoding the block code in one dimension followed by a second decoding in the other dimension. This makes the complexity of the product code decoder low for such a large code length.

5.2.2 Parallel Concatenated Block Codes.

Two-dimensional parallel turbo encoder is shown in Figure 5.3. Information bits are encoded twice by supplying the original information and its interleaved version to encoders 1 and 2, respectively. Figure 5.4 shows a 2D parallel concatenated block code where information is the 2-dimensional array and interleaver used is a block interleaver. The code looks similar to the serial concatenated code and can be considered as the product code without parity on parity.

5.3. Iterative Decoding of Concatenated Block Codes

In this section, two topics will be discussed. First, the ideas of serial and parallel iterative decoding are presented. Then, the details of the implementation of the

SISO decoding using the algebraic-based or augmented list decoding algorithm and trellis-based decoding algorithm are explained.

5.3.1 Serial Iterative Decoding

There are two SISO decoders, the Inner DEC and Outer DEC. The outer SISO decoder provides both $LLRs$ of the information and code symbols, not just the $LLRs$ of information symbols. The inner decoder needs to estimate $LLRs$ for the information symbols only. Figure 5.5 shows the serial turbo decoders where the L denotes LLR and the subscript e denotes the extrinsic information and the subscripts i and o denote the input to and output from a SISO decoder, respectively.

In the first iteration, the inner received symbols are fed to the inner decoder, i.e., Inner DEC, where the soft-output of the information part of an inner code, $L_{eo}(\hat{\tilde{b}})$, is obtained and de-interleaved. Then, it is passed to the Outer DEC as a priori value, $L_{ei}(\hat{b})$ for the code symbols of the outer code. The Outer DEC processes the outer received symbols with the a priori value and produces the extrinsic information of the code symbols of the outer code $L_{eo}(\hat{b})$. This extrinsic information is then interleaved and fed back to the Inner DEC as the priori value of information symbols. In the last iteration, the soft-output of the information symbols $L_o(\hat{a})$ are calculated at the Outer DEC and the signs of the soft values are the final decisions.

Note the difference between the serial iterative decoding presented here and the one used for more typical serial concatenated scheme [11] (see Chapter 2). Here, both decoders have access to the received signal, while in [11], only the inner code has direct access to the received signal and the outer decoder works only based on the information it receives form the inner decoder.

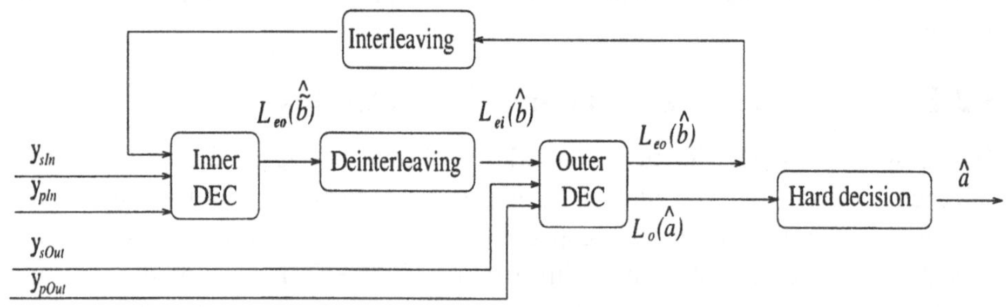

Figure 5.5. Serial Iterative Decoder

5.3.2 Parallel Iterative Decoding

Figure 5.6 shows the parallel iterative decoding of a turbo code. The received parity sequence Y_p is demultiplexed into Y_{1p} and Y_{2p}, which are the received

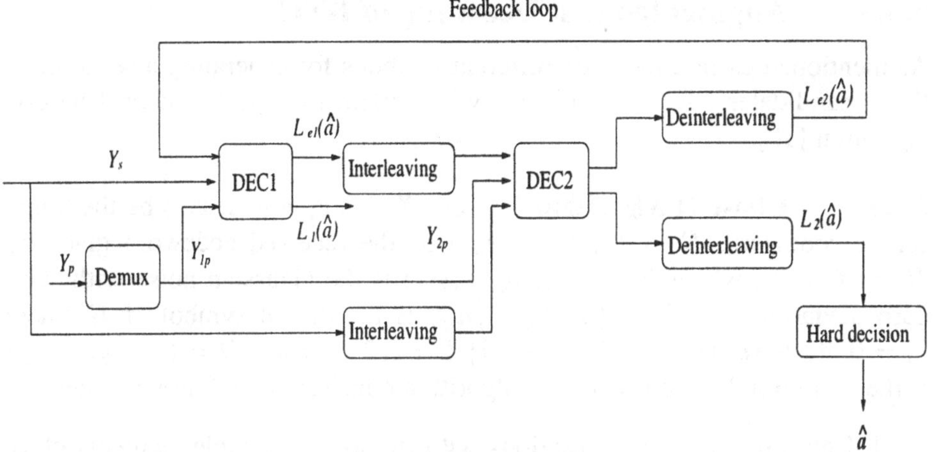

Figure 5.6. Parallel Iterative Decoder

parity subsequences for DEC1 and DEC2 respectively. DEC1 performs MAP decoding on the received information, Y_s, the received parity sequence, Y_{1p}, and the *a priori value*, $L(a)$ where the a priori value $L(a) = 0$ in the first decoding step. Then DEC1 provides the extrinsic information, $L_{e1}(\hat{a})$ which is fed to DEC2 as a priori value. Similarly, an interleaved version of the noise corrupted sequence as well as the a priori value obtained from DEC1 are fed into the MAP decoder, DEC2, where it provides the extrinsic value, $L_{e2}(\hat{a})$. Keep in mind that DEC1 and DEC2 operate over non-interleaved and interleaved versions of the received sequence, respectively. Thus, the input of DEC1 has to be de-interleaved if it is needed before being fed to the decoder. On the other hand, the inverse operation is applied to the input of DEC2. The soft-output provided by DEC1 and DEC2 for the information bit, a_k at iteration $t \geq 1$ is given by the relation,

$$\begin{aligned} L_1^{(t)}(a_k) &= L_c \cdot y_k + L_{e2}^{(t-1)}(a_k) + L_{e1}^{(t)}(a_k) \\ L_2^{(t)}(a_k) &= L_c \cdot y_k + L_{e1}^{(t-1)}(a_k) + L_{e2}^{(t)}(a_k) \end{aligned} \tag{5.1}$$

where $L_c = 4 \cdot a \cdot \frac{E_s}{N_o}$ is called the *reliability value* of the channel, a is the fading attenuation. For the Gaussian channel a equals to one and $\frac{E_s}{N_o}$ is the channel SNR estimated at the receiver.

For the final iteration the decision is obtained from the sign of de-interleaved soft-output of DEC2.

5.3.3 Augmented List Decoding of BTC

As mentioned earlier, there are different methods for generating a set of most likely candidates. In this book, we will explain the one based on Chase-II algorithm [10].

5.3.3.1 Chase-II Algorithm. Let $X = (x_1, x_2, ..., x_n)$ be the transmitted codeword, $R = (r_1, r_2, ..., r_n)$ be the received codeword given by $R = X + N$, where $N = (n_1, n_2, ..., n_n)$ is the Gaussian noise with standard deviation σ. $H = (h_1, h_2, ..., h_n)$ is the digital symbol of R where $h_l = 0.5(1+sign(r_l))$ and $h_l \in \{0, 1\}$; $l = 1, 2, ..., n$ and $D = (d_1, d_2, ..., d_n)$ is the optimum decision. Chase-II algorithm consists in the following steps.

1 Determine $p = \left\lfloor \frac{\delta}{2} \right\rfloor$ positions with the lowest reliable elements of H using R, where δ is the minimum distance of the code. In stationary AWGN channel, the normalized reliability of element h_j is given by $|r_j|$.

2 Form the set of error patterns E, the required error patterns defined as any n-bit binary sequence with $\left\lfloor \frac{\delta}{2} \right\rfloor$ 1's located in the positions found from step 1. So there are $2^{\lfloor \frac{\delta}{2} \rfloor}$ possible test patterns, including the all-zero pattern.

3 Form the set of test patterns $T = H \oplus E$, where $\oplus$ denote modulo 2 addition operation.

4 Decode all test patterns using an algebraic decoder and keep valid codewords in set C.

5 Make decision by using the decision rule

$$D = C^i \;\; if \;\; \left|R - C^i\right|^2 \leq \left|R - C^l\right|^2 \forall l \in [1, 2^{\lfloor \frac{\delta}{2} \rfloor}], \;\; l \neq i \tag{5.2}$$

where $C^i = (c_1^i, ..., c_l^i, ..., c_n^i)$ is the ith codeword of **C**. The square Euclidean distance between R and C^i is defined as

$$\left|R - C^i\right|^2 = \sum_{l=1}^{n} \left(r_l - c_l^i\right)^2 \tag{5.3}$$

By applying Chase algorithm to each row and column of the product codes, the near ML decoding of product codes is found. However, it is important to note that Chase algorithm does not guarantee to provide the most likely codeword. The reason is that it does not perform full search over all valid codewords. Therefore, this algorithm is sub-optimal.

5.3.3.2 Example of Chase Algorithm.

For better understanding of the decoding algorithm, we give an example of the Chase-II algorithm using actual information from a simulation run. In this case, we consider $p = 4$ least reliable binary elements.

R = (1.27,2.22,-0.78,-1.08,-0.99,0.48,**0.15**,-1.22,**0.21**,0.79,**-0.17**,2.33,**-0.41**,1.38,0.71)
H = (1,1,0,0,0,1,**1**,0,**1**,1,**0**,1,**0**,1,1)

2^p Error Patterns (E):

E^1	(0,0,0,0,0,0,**0**,0,**0**,0,**0**,0,**0**,0,0)	E^9	(0,0,0,0,0,0,**1**,0,**0**,0,**0**,0,**0**,0,0)
E^2	(0,0,0,0,0,0,**0**,0,**0**,0,**0**,0,**1**,0,0)	E^{10}	(0,0,0,0,0,0,**1**,0,**0**,0,**0**,0,**1**,0,0)
E^3	(0,0,0,0,0,0,**0**,0,**0**,0,**1**,0,**0**,0,0)	E^{11}	(0,0,0,0,0,0,**1**,0,**0**,0,**1**,0,**0**,0,0)
E^4	(0,0,0,0,0,0,**0**,0,**0**,0,**1**,0,**1**,0,0)	E^{12}	(0,0,0,0,0,0,**1**,0,**0**,0,**1**,0,**1**,0,0)
E^5	(0,0,0,0,0,0,**0**,0,**1**,0,**0**,0,**0**,0,0)	E^{13}	(0,0,0,0,0,0,**0**,0,**1**,0,**0**,0,**0**,0,0)
E^6	(0,0,0,0,0,0,**0**,0,**1**,0,**0**,0,**1**,0,0)	E^{14}	(0,0,0,0,0,0,**1**,0,**1**,0,**0**,0,**1**,0,0)
E^7	(0,0,0,0,0,0,**0**,0,**1**,0,**1**,0,**0**,0,0)	E^{15}	(0,0,0,0,0,0,**1**,0,**1**,0,**1**,0,**0**,0,0)
E^8	(0,0,0,0,0,0,**0**,0,**1**,0,**1**,0,**1**,0,0)	E^{16}	(0,0,0,0,0,0,**1**,0,**1**,0,**1**,0,**1**,0,0)

2^p Test Sequences $(T = H \bigoplus E)$:

T^1	(1,1,0,0,0,1,**1**,0,**1**,1,**0**,1,**0**,1,1)	T^9	(1,1,0,0,0,1,**0**,0,**1**,1,**0**,1,**0**,1,1)
T^2	(1,1,0,0,0,1,**1**,0,**1**,1,**0**,1,**0**,1,1)	T^{10}	(1,1,0,0,0,1,**0**,0,**1**,1,**0**,1,**1**,1,1)
T^3	(1,1,0,0,0,1,**1**,0,**1**,1,**1**,1,**0**,1,1)	T^{11}	(1,1,0,0,0,1,**0**,0**1**,1,**1**,1,**0**,1,1)
T^4	(1,1,0,0,0,1,**1**,0,**1**,1,**1**,1,**1**,1,1)	T^{12}	(1,1,0,0,0,1,**0**,0,**1**,1,**1**,1,**1**,1,1)
T^5	(1,1,0,0,0,1,**1**,0,**0**,1,**0**,1,**0**,1,1)	T^{13}	(1,1,0,0,0,1,**0**,0,**0**,1,**0**,1,**0**,1,1)
T^6	(1,1,0,0,0,1,**1**,0,**0**,1,**0**,1,**1**,1,1)	T^{14}	(1,1,0,0,0,1,**0**,0,**0**,1,**0**,1,**1**,1,1)
T^7	(1,1,0,0,0,1,**1**,0,**0**,1,**1**,1,**0**,1,1)	T^{15}	(1,1,0,0,0,1,**0**,0,**0**,1,**1**,1,**0**,1,1)
T^8	(1,1,0,0,0,1,**1**,0,**0**,1,**1**,1,**1**,1,1)	T^{16}	(1,1,0,0,0,1,**0**,0,**0**,1,**1**,1,**1**,1,1)

Decoding T using algebraic decoder

C^1	(1,1,0,0,0,1,1,0,1,1,0,1,0,1,1)	C^9	(1,1,0,0,0,1,1,0,1,1,0,1,0,1,1)
C^2	(1,1,0,0,0,1,1,0,1,1,0,1,0,1,1)	C^{10}	(1,1,0,0,0,1,1,0,1,1,0,1,0,1,1)
C^3	(1,1,0,0,0,1,1,0,1,1,0,1,0,1,1)	C^{11}	(1,1,0,0,0,1,1,0,1,1,0,1,0,1,1)
C^4	(1,1,0,0,0,1,1,0,1,1,1,1,1,0,0)	C^{12}	(1,1,0,0,0,1,1,0,1,1,0,1,0,1,1)
C^5	(1,1,0,0,0,1,1,0,1,1,0,1,0,1,1)	C^{13}	(1,1,0,0,0,1,1,0,1,1,0,1,0,1,1)
C^6	(x,x,x,x,x,x,x,x,x,x,x,x,x,x,x)	C^{14}	(1,1,0,0,0,1,1,0,1,1,0,1,0,1,1)
C^7	(1,1,1,0,0,1,1,0,0,0,1,1,0,1,1)	C^{15}	(1,1,0,0,0,1,1,0,1,1,0,1,0,1,1)
C^8	(x,x,x,x,x,x,x,x,x,x,x,x,x,x,x)	C^{16}	(1,1,0,0,1,1,0,0,0,1,1,0,1,1,1)

where (x,x,x,x,x,x,x,x,x,x,x,x,x,x,x) represents the invalid codeword which results when the number of errors exceeds the error correcting capability of the code. In this case, the code is BCH code with error correcting capability of one. Then, the calculation of the square Euclidean distance between R and $C^l \in \mathbf{C}$, $l \in [1, 2^{\lfloor \frac{\delta}{2} \rfloor}]$ is performed in order to find the decision D

$Dist^1$	6.96	$Dist^9$	6.96
$Dist^2$	6.96	$Dist^{10}$	6.96
$Dist^3$	6.96	$Dist^{11}$	6.96
$Dist^4$	17.04	$Dist^{12}$	6.96
$Dist^5$	6.96	$Dist^{13}$	6.96
$Dist^6$	X	$Dist^{14}$	6.96
$Dist^7$	14.15	$Dist^{15}$	6.96
$Dist^8$	X	$Dist^{16}$	23.4

Decision codeword, D at distance of 6.96 from R is (1, 1, 0, 0, 0, 1, 1, 0, 1, 1, 0, 1, 0, 1, 1).

5.3.3.3 Reliability of Decision D.

The reliability of the decision bit, d_j defined by the log-likelihood ratio(LLR) of the transmitted bit x_j is given by,

$$\Lambda(d_j) = \ln\left(\frac{Pr\{x_j = +1|R\}}{Pr\{x_j = -1|R\}}\right) \tag{5.4}$$

where

$$Pr\{x_j = +1|R\} = \sum_{C^i \in S_j^{+1}} Pr\{X = C^i|R\} \tag{5.5}$$

$$Pr\{x_j = -1|R\} = \sum_{C^i \in S_j^{-1}} Pr\{X = C^i|R\} \tag{5.6}$$

where $S_j^{+1}$and S_j^{-1} are the sets of codewords in C such that $c_j^i = +1$ and $c_j^i = -1$ respectively. It is assumed that each codeword is transmitted with equal probability. Thus, from the Bayes rule, the LLR of decision d_j is given as,

$$\Lambda(d_j) = \ln\left(\frac{\sum_{C^i \in S_j^{+1}} p\{R \mid X = C^i\}}{\sum_{C^i \in S_j^{-1}} p\{R \mid X = C^i\}}\right) \tag{5.7}$$

where the conditional probability density under Gaussian channel assumption is given by

$$p\left\{R \mid X = C^i\right\} = \left(\frac{1}{\sqrt{2\pi}\sigma}\right)^n \exp\left(-\frac{|R - C^i|^2}{2\sigma^2}\right) \tag{5.8}$$

Thus, the LLR of decision d_j can be written as

$$\Lambda(d_j) = \frac{1}{2\sigma^2}\left(\left|R - C^{-1(j)}\right|^2 - \left|R - C^{+1(j)}\right|^2\right) + \ln\left(\frac{\sum_i A_i}{\sum_i B_i}\right) \tag{5.9}$$

where

$$A_i = \exp\left(\frac{\left|R - C^{+1(j)}\right|^2 - \left|R - C^i\right|^2}{2\sigma^2}\right) \leq 1 \tag{5.10}$$

with $C^i \in S^{+1(j)}$

$$B_i = \exp\left(\frac{\left|R - C^{-1(j)}\right|^2 - \left|R - C^i\right|^2}{2\sigma^2}\right) \leq 1 \tag{5.11}$$

with $C^i \in S^{-1(j)}$

The sums $\sum_i A_i$ and $\sum_i B_i$ both tend to zero at high signal to noise ratios, i.e., when σ tends to zero. Therefore, the LLR can be approximated by omitting the second expression, and equals to

$$\Lambda'(d_j) = \frac{1}{2\sigma^2}\left(\left|R - C^{-1(j)}\right|^2 - \left|R - C^{+1(j)}\right|^2\right) \tag{5.12}$$

where $C^{+1(j)}$ and $C^{-1(j)}$ are the codewords closest to R in $S_j^{+1}$and S_j^{-1}, respectively.

Using Equation (5.3), the the LLR is,

$$\Lambda'(d_j) = \frac{2}{\sigma^2}\left(r_j + \sum_{l=1, l\neq j}^{n} r_l c_l^{+1(j)} p_l\right) \tag{5.13}$$

where

$$p_l = \begin{cases} 0, & if \quad c_l^{+1(j)} = c_l^{-1(j)} \\ 1, & if \quad c_l^{+1(j)} \neq c_l^{-1(j)} \end{cases} \tag{5.14}$$

Normalizing $\Lambda'(d_j)$ by $\frac{2}{\sigma^2}$, we get,

$$r'_j = r_j + w_j \tag{5.15}$$

with

$$w_j = \sum_{l=1,l\neq j}^{n} r_l c_l^{+1(j)} p_l \tag{5.16}$$

5.3.3.4 Computing the Soft Decision at the Output of the Soft-input Decoder. In order to compute the reliability of decision bit, d_j , two codewords are required. One is the decision codeword D and the other one is a competing codeword, B. The competing codeword for D is the codeword B with $b_j \neq d_j$ with the minimum Euclidean distance from R, i.e., the bit at position j is complement to that of D.

Since we use the Chase algorithm in order to find the set **C** of possible candidates for the optimal decision D, we also use that set to find B. However, B is not always found so in [10], two ways are proposed for calculating soft output.

Case 1: when B is found,

$$r_j' = \left(\frac{|R-B|^2 - |R-D|^2}{4}\right) d_j \tag{5.17}$$

Case 2: when B is not found,

$$r_j' = \beta \times d_j \;\; with \;\; \beta \geq 0 \tag{5.18}$$

where β is the reliability of decision bit when the competing codeword is not found and is given as

$$\beta \approx \left| \ln \left(\frac{Pr\{d_j = x_j\}}{Pr\{d_j \neq x_j\}} \right) \right| \tag{5.19}$$

In [10], the reliability factor, β is given as a function of BER or the number of decoding step, t as

$$\beta(t) = [0.2, 0.4, 0.6, 0.8, 1.0, 1.0, 1.0, 1.0] \tag{5.20}$$

In a 2D product code, two decoding steps, i.e. decoding horizontally and vertically , are considered as one decoding iteration. It is noted that when competing codeword, B is not found, the accuracy of the reliability value of decision bit is not crucial. It is because the probability of the correct decision is high since B is far from the received sequence B, thus the average value of the reliability is sufficient.

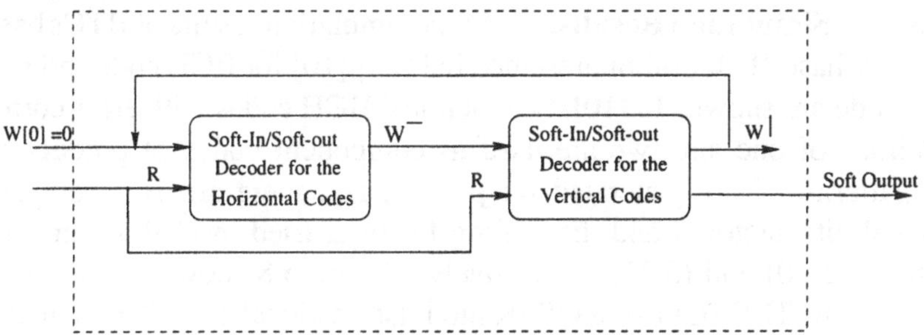

Figure 5.7. The Tubo Decoding Process

5.3.3.5 Iterative Decoding of Product Codes. In a product code, we decode each row of product code, horizontal decoding, and calculate the soft output for each coded bit. Extrinsic information of rows W^- can be calculated by,

$$W^- = R^-_{output} - R^-_{input} \tag{5.21}$$

W^- is averaged and normalized to one to obtain independence between the decoding steps. Then, $W^-$is passed to the next decoding step providing soft-input to the vertical decoder as follow:

$$R^|_{input} = R + \alpha W^- \tag{5.22}$$

where α is a scaling factor used to reduce the effect of extrinsic information which is not reliable at the very first iteration. In addition, it is introduced due to the fact that standard deviations of the samples of R and W are different. Value of α increase as the number of decoding steps grows. The values of α for different iterations t, are

$$\alpha(t) = [0.0, 0.2, 0.3, 0.5, 0.7, 0.9, 1.0, 1.0] \tag{5.23}$$

We will get extrinsic information of the columns $W^|$ in the same way as W^-. Iterative or turbo decoding process of BTC is shown in Figure 5.7.

5.3.3.6 Simulation Results. Some simulation results of BTCs based on modified Chase-II algorithm presented in [128],[10] for BCH code and in [130] for RS code are shown. In [10], two identical BCH codes with error correcting capabilities of one and two are used as component codes of product codes. Error patterns, E are generated for $p = 4$, i.e., there are 16 error patterns. The reliability factor β and the scaling factor α used in [10] are given as in Equations (5.20) and (5.23), respectively. Figure 5.8 shows the performance of different BCH-BTCs using QPSK modulation signal over Gaussian channel after 4 iterations. Figure 5.8 shows that the slope of the (BER, $\frac{E_b}{N_o}$) curves increases with an increase in either n, k or the minimum distance.

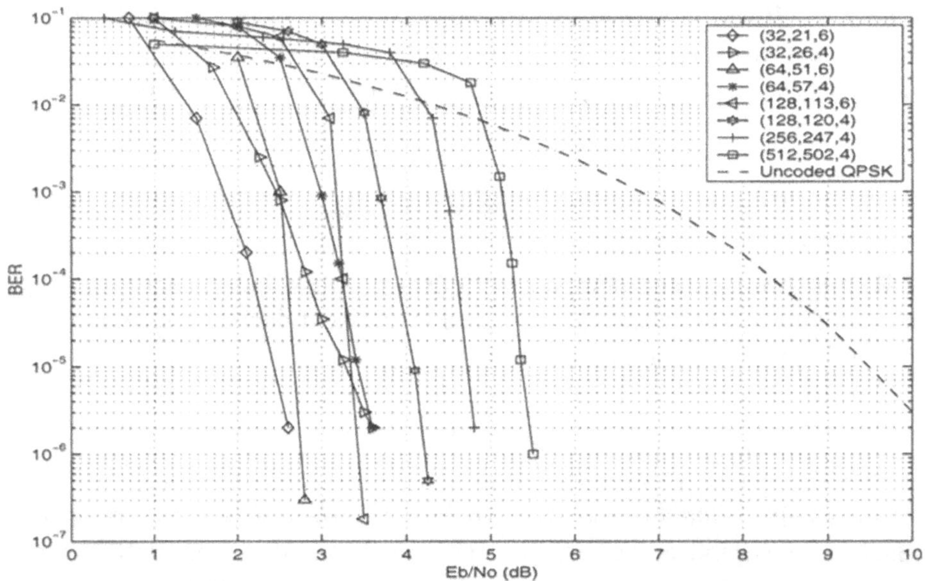

Figure 5.8. Performance of BCH-TPCs using QPSK Modulation after 4 Iterations over AWGN Channel

Figure 5.9 shows the $\frac{E_b}{N_o}$ required to achieve a BER of 10^{-5} for different BTCs after 4 iterations. These are compared with the Shannon's limit and cut-off rate for the binary input Gaussian channel. It is shown that almost all presented codes achieve the performance with code rates exceeding the cut-off rates. Furthermore, the BTC $(512, 502, 4)^2$ performs less than 0.8 dB from the Shannon's limit for an AWGN channel with binary input while operating near channel capacity $R/C > 0.98$.

In [10], the performance of BCH-BTCs over Rayleigh fading channel after 4 iterations is investigated and presented in Figure 5.10. It is noted that the slope of the BER curves are as steep as in the case of Gaussian channel with a shift of about 4.6 dB to the right. In [130], the performance of RS-BTCs is presented.

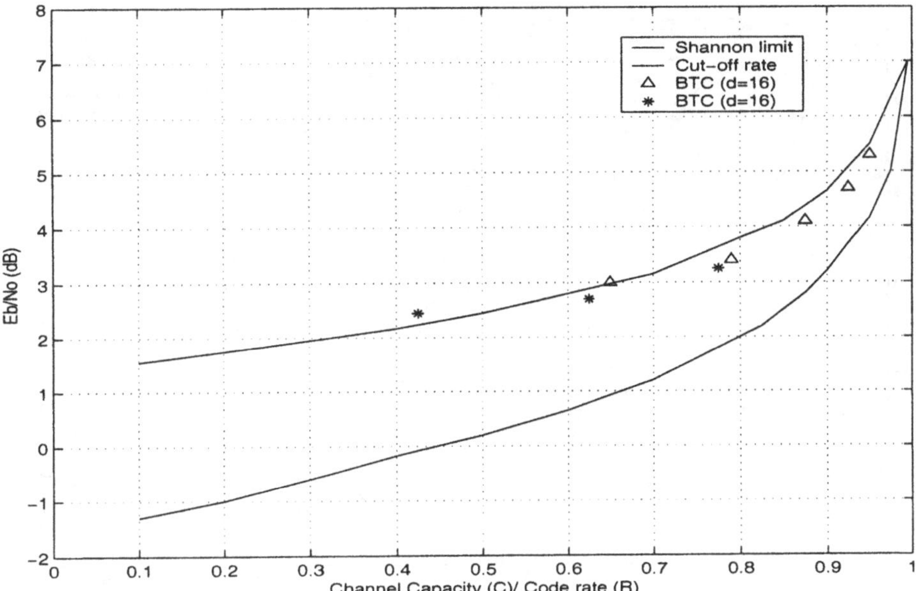

Figure 5.9. Performance Comparison of BCH-TPCs after 4 Iterations over AWGN Channel and Theoretical Limits for Gaussian Channel with Binary Input

The summary of the performance of different RS-BTCs at BER of 10^{-5} after 4 iterations is shown in Table 5.2. It is noted in [130] that RS-BTCs are not as powerful as BCH-BTCs.

RS Product Code $C_1 = C_2$	*Rate*	d_{min}	Eb/No dB @BER 10^{-5}	*Shannon's limit (dB)*
$(15, 12, 4)$	0.64	16	4.10	0.47
$(31, 28, 4)$	0.81	16	4.42	1.07
$(31, 26, 6)$	0.70	36	4.10	0.68
$(31, 24, 8)$	0.60	64	4.06	0.33
$(63, 60, 4)$	0.91	16	5.06	1.43
$(63, 58, 6)$	0.84	36	4.63	1.20
$(63, 56, 8)$	0.79	64	4.40	1.00

Table 5.2. Performance of RS-TPCs after 4 Iterations on AWGN Channel

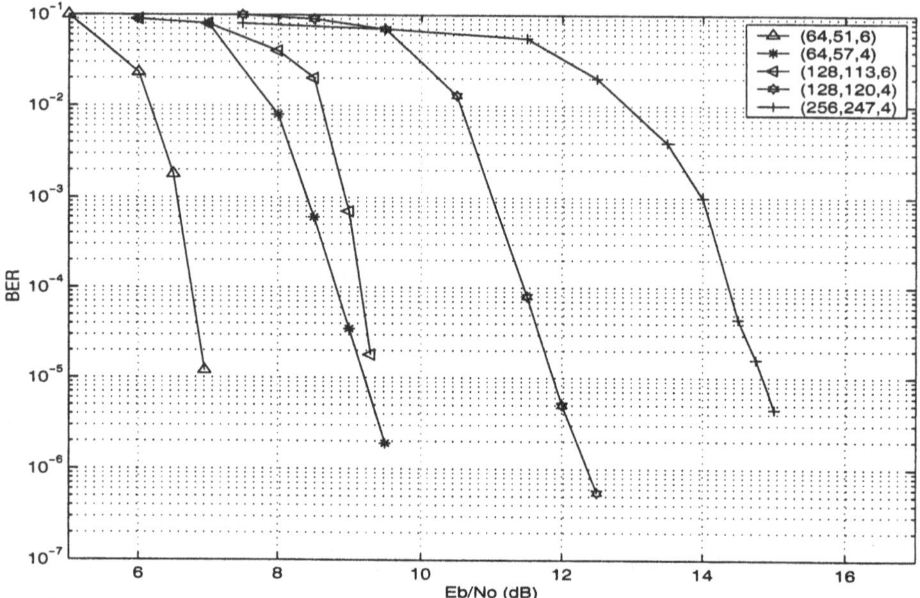

Figure 5.10. Performance of BCH-TPCs using QPSK Modulation after 4 Iterations over Rayleigh Fading Channel

5.3.4 Trellis-based Decoding of BTC

In this section, trellis-based *Maximum a posteriori Probability* (MAP) algorithm is presented. The following mathematical derivations are presented in [7] and is similar to the results for the convolutional turbo codes in Chapter 2.

5.3.4.1 MAP Algorithm. The binary trellis of a block code is shown in Figure 5.11. The coded bit at the time m, x_m is the label of the branch that connects from time $m-1$ to time m. The trellis states at times $m-1$ and m are s' and s, respectively.

The soft-output of a given information bit, x_m from the MAP decoder is defined as the conditional a posteriori log-likelihood ratio when the sequence $\overrightarrow{y}$ is received. The soft-output is given by

$$L(\widehat{x}_m) = L(x_m \mid \overrightarrow{y}) = \ln \frac{Pr(x_m = +1 \mid \overrightarrow{y})}{Pr(x_m = -1 \mid \overrightarrow{y})} = \ln \frac{\sum_{(s',s),x_m=+1} p(s', s, \overrightarrow{y})}{\sum_{(s',s),x_m=-1} p(s', s, \overrightarrow{y})} \tag{5.24}$$

In the memoryless channel, the joint probability $p(s', s, \overrightarrow{y})$ can be written as follows:

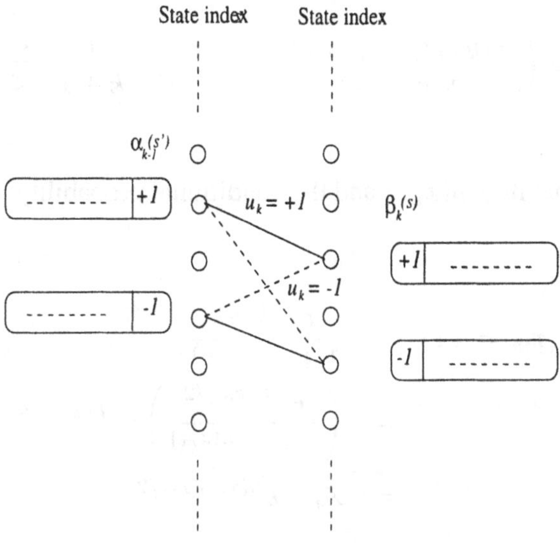

Figure 5.11. Trellis Structure of a Systematic Block Code

$$\begin{aligned} p(s', s, \vec{y}) &= p(s', \vec{y}_{j<m}) \cdot p(s, \vec{y}_m \mid s') \cdot p(\vec{y}_{j>m} \mid s) \\ &= p(s', \vec{y}_{j<m}) \cdot P(s \mid s') \cdot p(\vec{y}_m \mid s', s) \cdot p(\vec{y}_{j>m} \mid s) \\ &= \alpha_{m-1}(s') \cdot \gamma_m(s', s) \cdot \beta_m(s) \end{aligned} \tag{5.25}$$

where $\vec{y}_{j<m}$ represents the portion of the received sequence from bit 0 up to bit $m-1$. Similarly, the received sequence from bit m up to bit $n-1$ is denoted by $\vec{y}_{j>m}$. $\alpha_m(s)$ and $\beta_{m-1}(s')$ are defined as the forward and backward recursions of the MAP decoding, respectively.

$$\alpha_m(s) = \sum_{s'} \gamma_m(s', s) \cdot \alpha_{m-1}(s') \tag{5.26}$$

$$\beta_{m-1}(s') = \sum_{s} \gamma_m(s', s) \cdot \beta_m(s) \tag{5.27}$$

where $\alpha_0(0) = 1$ and $\beta_n(0) = 1$. The branch transition probability is given by

$$\gamma_m(s', s) = P(s|s') \cdot p(y_m|s', s) = p(x_m; y_m) \tag{5.28}$$

We assume that the information bits are statistically independent. In an (n, k) systematic block codes, the transition probability is given by,

$$p\left(x_m; y_m\right) = \begin{cases} p(y_m|x_m) \cdot p(x_m) & 1 \leq m \leq k \\ p(y_m|x_m) & k+1 \leq m \leq n \end{cases} \tag{5.29}$$

A priori probability, $p(x_m)$, and the conditional probability, $p(y_m \mid x_m)$, are given as,

$$\begin{aligned} p(x_m = \pm 1) &= \frac{e^{\pm L(x_m)}}{1 + e^{\pm L(x_m)}} \\ &= \left(\frac{e^{-L(x_m)/2}}{1 + e^{-L(x_m)}}\right) \cdot e^{L(x_m)\cdot x_m/2} \\ &= A_m \cdot e^{L(x_m)\cdot x_m/2} \end{aligned} \tag{5.30}$$

$$\begin{aligned} p(y_m \mid x_m = \pm 1) &= \left(\frac{P(x_m) \cdot (1 + e^{-L(x_m)}) \cdot e^{-L_c \cdot y_m/2}}{1 + e^{-(L(x_m) + L_c \cdot y_m)}}\right) \cdot e^{L_c \cdot y_m \cdot x_m/2} \\ &= B_m \cdot e^{L_c \cdot y_m \cdot x_m/2} \end{aligned} \tag{5.31}$$

The log-likelihood ratio associated with $p(x_m; y_m)$ can be written as

$$L\left(x_m; y_m\right) = \begin{cases} L_c y_m + L(x_m) & 1 \leq m \leq k \\ L_c y_m & k+1 \leq m \leq n \end{cases} \tag{5.32}$$

A_m and B_m in Equations (5.30) and (5.31) are equal for all transitions from time $m-1$ to time m and can be omitted due to the ratio in Equation (5.24). As a result, a simplified version of branch transition probability can be expressed as follows:

$$\gamma_m(s', s) = e^{\left(\frac{1}{2} x_m (L_c \cdot y_m + L(x_m))\right)} = e^{(L(x_m; y_m)\cdot x_m/2)} \tag{5.33}$$

In a systematic block code, a priori probability $L(x_m)$ is equal to zero if x_m is a parity bit.

5.3.4.2 Soft-Output Calculation. Soft-outputs from the log-MAP and Max-log-MAP decoding are given in this section

- The optimal soft-output using the log-MAP decoder can be written as

$$L(\widehat{x}_m) = L_c y_m + L(u_m) + \frac{\sum_{(s',s), u_m=+1} \alpha_{m-1}(s') \cdot \beta_m(s)}{\sum_{(s',s), u_m=-1} \alpha_{m-1}(s') \cdot \beta_m(s)} \tag{5.34}$$

where forward and backward recursions are given in Equations (5.26) and (5.27). The last term in Equation (5.34) is the extrinsic information

- The sub-optimum soft-output using Max log-MAP for systematic block codes is the approximated version of log-MAP algorithm. Using the approximation $\ln(e^{a_1} + e^{a_2} + \cdots + e^{a_m}) \approx \max_{0 \le i \le m} a_i$, Equation (5.34) can be approximated and the log-likelihood ratio is given by

$$\begin{aligned} L_{log-MAP}(\widehat{x}_m) &= L_c \cdot y_m + L(u_m) \qquad (5.35) \\ &+max_{(s',s),u_m=+1}\left(\ln(\alpha_{m-1}(s')) + \ln(\beta_m(s))\right) \\ &-max_{(s',s),u_m=-1}\left(\ln(\alpha_{m-1}(s')) + \ln(\beta_m(s))\right) \end{aligned}$$

with

$$\ln(\alpha_m(s)) = max_{s'}\left(\ln(\alpha_{m-1}(s')) + \frac{1}{2}L(x_m; y_m) \cdot x_m\right) \qquad (5.36)$$

$$\ln(\beta_{m-1}(s')) = max_s\left(\ln(\beta_m(s)) + \frac{1}{2}L(x_m; y_m) \cdot x_m\right) \qquad (5.37)$$

The soft-output can also be calculated from the modified trellis and the trellis of the dual code. For further detail may be found in [7].

5.4. Summary

This chapter presented the idea of block turbo codes, including the way they are encoded and decoded. Two main decoding approaches were presented, namely, the algebraic decoding and Trellis-based decoding. The details of the algebraic decoding, i.e., modified Chase-II algorithm were presented, along with an example. Some simulation results from the literatures of BTCs using algebraic-based iterative decoding were also presented. It was shown that BCH-BTCs perform well at high code rates, i.e., $\frac{R}{C} \approx 1$.

Chapter 6

REED-MULLER CODES AND REED-MULLER TURBO CODES

6.1. Introduction

In this chapter, *Reed-Muller* (RM) codes and Reed-Muller turbo codes are discussed. We present definition and properties of RM codes. It was stated in Chapter 5 that there are two approaches used in MAP decoding of block codes. One is based on the list decoding algorithm, i.e. the modified Chase-II algorithm[128] . It is noted that list decoding is sub-optimal because it does not perform full search over all valid codewords. Trellis-based decoding algorithm, however, is optimal. Our focus will be on the latter. The issue of constructing the trellis of block codes and, particularly their *minimal trellis* representation, is then considered. Finally, the details of RM-turbo codes, their encoder and decoder are presented. The chapter is organized as follows.

In the second section, the definition and properties of Reed-Muller codes are presented. In the third section, we present the definitions related to the trellis diagram of block codes. Then the construction of the trellis diagram of a linear block code using BCJR [57] and Massey algorithm [145] is discussed. In particular, the construction of trellis diagram of a RM code is presented. Then, turbo encoder and decoder will be presented. The presented encoder is a parallel concatenated code constructed from two elementary encoders with an interleaver between them. The decoder is an iterative MAP decoding algorithm. Then, the system model used for the simulation purpose is given. The simulation results of RM turbo codes on Additive White Gaussian Noise (AWGN) and Rayleigh-fading channels are shown. Modified RM turbo codes, the shortened codes, are then investigated for use in satellite ATM application. The design of shortened RM-turbo codes with different shortening patterns will be discussed. Also the performance of the shortened version of the proposed coding scheme is investigated. It is shown that some shortened patterns obtain *Unequal Error Protection* (UEP) property. A UEP code is more suitable for the structure of ATM cell since cell-header is more important than its payload. The coding scheme presented in this chapter is the result of the research work of the authors.

6.2. Reed-Muller Codes.

Reed-Muller codes can be defined in terms of Boolean functions. In order to define codes of length $n = 2^m$, we need m basis vectors of length 2^m, denoted as $\overrightarrow{v_1}, \overrightarrow{v_2}, ..., \overrightarrow{v_m}$ whose elements take the values 0 and 1. Let $\overrightarrow{V} = (\overrightarrow{v_1}, \overrightarrow{v_2}, ..., \overrightarrow{v_m})$ range over V^m, the set of all binary $m - tuples$ in increasing or decreasing order. Let $\overrightarrow{a} \cdot \overrightarrow{b}$ be the *Boolean product* of vectors $\overrightarrow{a}$ and $\overrightarrow{b}$ where $\overrightarrow{a} = (a_1, a_2, ..., a_n)$, $\overrightarrow{b} = (b_1, b_2, ..., b_n)$ and $'\cdot'$ is the AND operation.

$$\overrightarrow{a} \cdot \overrightarrow{b} = (a_1 \cdot b_1, a_2 \cdot b_2, ..., a_n \cdot b_n) \tag{6.1}$$

For simplicity, $\overrightarrow{a} \cdot \overrightarrow{b}$ is denoted by $\overrightarrow{ab}$. The vector obtained from a Boolean product of l vectors is said to be a polynomial of degree l. Boolean function, $f(\overrightarrow{V}) = f(\overrightarrow{v_1}, \overrightarrow{v_2}, ..., \overrightarrow{v_m})$, is defined as any function resulting from the AND operation of its arguments. The following are the definitions and code parameters of the RM code.

Definition 1: Let $0 \leq r \leq m$, the binary Reed-Muller code $\Re(r, m)$ of order r and length 2^m consists of the vectors $\overrightarrow{f}$ associated with all Boolean functions f, that are polynomials of degree less than or equal to r in m variables

Code parameters :

- The code length, n is 2^m.
- The dimension or the message length, k, of $\Re(r, m)$ is defined as:

$$k = 1 + \binom{m}{1} + \binom{m}{2} + ... + \binom{m}{r} \tag{6.2}$$

- The minimum distance, d_{min} of $\Re(r, m)$ is 2^{m-r}.

The generator matrix, G of $\Re(r, m)$ with order r and length 2^m consists of vectors with polynomial degree less than or equal to r and can be constructed as follows:

$$G = \begin{bmatrix} \overrightarrow{1} \\ ---- \\ \overrightarrow{v_m} \\ \overrightarrow{v_{m-1}} \\ \vdots \\ \overrightarrow{v_1} \\ ---- \\ \overrightarrow{v_{m-1}v_m} \\ \overrightarrow{v_{m-2}v_m} \\ \vdots \\ \overrightarrow{v_1 v_2} \\ ---- \\ \vdots \\ ---- \\ \overrightarrow{v_{m-r+1}v_{m-r+2}\cdots v_m} \\ \overrightarrow{v_{m-r}v_{m-r+2}\cdots v_m} \\ \vdots \\ \overrightarrow{v_1 v_2 \cdots v_r} \end{bmatrix} \begin{matrix} ---> & degree\ 0 \\ ---> & degree\ 1 \\ ---> & degree\ 2 \\ \vdots & \vdots \\ ---> & degree\ r \end{matrix} \tag{6.3}$$

Consider the example of $\Re(2,4)$ code. The 4 basis vectors of length 16 are given by,

$$\begin{array}{lcllllllllllllllll} \overrightarrow{v_4} & = & 0 & 0 & 0 & 0 & 0 & 0 & 0 & 0 & 1 & 1 & 1 & 1 & 1 & 1 & 1 & 1 \\ \overrightarrow{v_3} & = & 0 & 0 & 0 & 0 & 1 & 1 & 1 & 1 & 0 & 0 & 0 & 0 & 1 & 1 & 1 & 1 \\ \overrightarrow{v_2} & = & 0 & 0 & 1 & 1 & 0 & 0 & 1 & 1 & 0 & 0 & 1 & 1 & 0 & 0 & 1 & 1 \\ \overrightarrow{v_1} & = & 0 & 1 & 0 & 1 & 0 & 1 & 0 & 1 & 0 & 1 & 0 & 1 & 0 & 1 & 0 & 1 \end{array} \tag{6.4}$$

where the 16 columns represented the binary 4-tuples in increasing order.

A generator matrix for this code is constructed from vectors obtained from all Boolean functions of the 4 basis vectors with polynomial degree less than or equal to 2.

$$G_{\Re(2,4)} = \left[\begin{array}{cccccccccccccccccc} \vec{v_0} & = & 1 & 1 & 1 & 1 & 1 & 1 & 1 & 1 & 1 & 1 & 1 & 1 & 1 & 1 & 1 & 1 \\ \vec{v_4} & = & 0 & 0 & 0 & 0 & 0 & 0 & 0 & 0 & 1 & 1 & 1 & 1 & 1 & 1 & 1 & 1 \\ \vec{v_3} & = & 0 & 0 & 0 & 0 & 1 & 1 & 1 & 1 & 0 & 0 & 0 & 0 & 1 & 1 & 1 & 1 \\ \vec{v_2} & = & 0 & 0 & 1 & 1 & 0 & 0 & 1 & 1 & 0 & 0 & 1 & 1 & 0 & 0 & 1 & 1 \\ \vec{v_1} & = & 0 & 1 & 0 & 1 & 0 & 1 & 0 & 1 & 0 & 1 & 0 & 1 & 0 & 1 & 0 & 1 \\ \overrightarrow{v_3 v_4} & = & 0 & 0 & 0 & 0 & 0 & 0 & 0 & 0 & 0 & 0 & 0 & 0 & 1 & 1 & 1 & 1 \\ \overrightarrow{v_2 v_4} & = & 0 & 0 & 0 & 0 & 0 & 0 & 0 & 0 & 0 & 0 & 1 & 1 & 0 & 0 & 1 & 1 \\ \overrightarrow{v_1 v_4} & = & 0 & 0 & 0 & 0 & 0 & 0 & 0 & 0 & 0 & 1 & 0 & 1 & 0 & 1 & 0 & 1 \\ \overrightarrow{v_2 v_3} & = & 0 & 0 & 0 & 0 & 0 & 0 & 1 & 1 & 0 & 0 & 0 & 0 & 0 & 0 & 1 & 1 \\ \overrightarrow{v_1 v_3} & = & 0 & 0 & 0 & 0 & 0 & 1 & 0 & 1 & 0 & 0 & 0 & 0 & 0 & 1 & 0 & 1 \\ \overrightarrow{v_1 v_2} & = & 0 & 0 & 0 & 1 & 0 & 0 & 0 & 1 & 0 & 0 & 0 & 1 & 0 & 0 & 0 & 1 \end{array} \right] \tag{6.5}$$

The dimension of the code, k equals to 11 because the constraint of the code; the degree of the polynomial of all Boolean functions has to be less than or equal to 2.

$$k = 11 = 1 + \binom{4}{1} + \binom{4}{2} \tag{6.6}$$

The code that is generated by the generator matrix constructed above is not in a systematic form. However, the RM code is a linear block code, so the generator matrix can be modified by using linear operations on its rows to make it a systematic-like code. We will describe this in the next section.

It is common to describe a code with parameters n and k, we will use the notation RM(n, k) instead of $\Re(r, m)$ in this book. For example, we denote the $\Re(2, 4)$ code discussed above as an RM(16,11) code.

6.3. Minimal Trellis for Linear Block Codes

In [137], the trellis diagram construction and the Maximum-Likelihood (ML) decoding of block codes were presented. One obvious factor that determines the complexity of a trellis-based decoder for a block code is the structure of its trellis (the number of states and branches). However, it has been found that there are many trellis representations for a given block code. Thus, one way to reduce the complexity of the decoder is to seek the "*Minimal Trellis*" which means the best-trellis representation in the sense of having the smallest number of states and branches than any other trellis-representations. Recently, there has been a lot of attention on the trellis structure of block codes [138]-[141]. As it is stated in [138], different trellis representations are obtained from different orderings (permutations) of the symbol positions of any given block code. To date, the problem of finding the minimal trellis of a block code obtained by any permutation has not been solved in general and has been stated to be an

NP-complete problem [143]. However, there are some codes whose minimal trellis is known. These include the RM code [144] and Goley code [138]

In this section, we first introduce some basic notations and definitions related to trellis representation of linear block codes [146]. Next, we will discuss the minimal trellis construction using BCJR [57] and Massey methods [145], [146]. The former trellis construction method is based on parity check matrix of the code, whereas the latter is based on the generator matrix. We will use Massey algorithm to construct trellis of RM codes in the next section. The reason is that the generator matrix of RM codes is obviously obtained, however, its parity check matrix is not. Moreover, Massey algorithm constructs the trellis diagram of a *systematic* linear block code .

6.3.1 Notations and Definitions

Some preliminary notations and definitions used to explain the trellis of block codes are defined as follows:

Definition 1: A trellis of depth $n, T = (S, B, L)$, is a directed graph of length n. It consists of three sets of elements as follows:

the states, S, the branches, B and the labels, L, where, each set can be decomposed into subsets as given below:

$$\begin{aligned} S &= S_0 \cup S_1 \cup \cdots \cup S_n \\ B &= B_0 \cup B_1 \cup \cdots \cup B_{n-1} \\ L &= L_0 \cup L_1 \cup \cdots \cup L_{n-1} \end{aligned} \tag{6.7}$$

At time i, a subset S_i consists of m_i states. The subsets S_o and S_n, each have only one state called the original state s_o and the final state s_f , respectively. At section i, each branch b_{ij} in B_i connects a state in S_i to a state in S_{i+1} with a branch-label l_{ij} in L_i.

Definition 2: A trellis of depth n, $T = (S, B, L)$ represents a linear block code C of length n, if the sequence of branch labels of each path uniquely corresponds to a codeword in C.

Definition 3: A trellis T for a code C is called minimal, if the number of states at each time $i \; : \; i = 0, 1, ..., n$ is minimal among all possible trellis representations of C.

Definition 4: Let $\overrightarrow{a} = (a_1, a_2, ..., a_n)$ be a non zero vector over $GF(q)$. $L(\overrightarrow{a})$ is called the left index of $\overrightarrow{a}$ and equals to the smallest index i such that $a_i \neq 0$.

Definition 5: Let $G = (\vec{g}_0, \vec{g}_1, ..., \vec{g}_{k-1})^T$ be a $k \times n$ matrix with k row vectors of length n, $\vec{g}_i \ : \ i = 0, 1, ..., k-1$ over $GF(q)$. G is said to be in a reduced echelon form if,

$$L(\vec{g}_0) < L(\vec{g}_1) < \cdots < L(\vec{g}_{k-1}) \qquad (6.8)$$

and k columns of G at positions $L(\vec{g}_i) \ : \ i = 0, 1, ..., k-1$ have weight one.

6.3.2 Minimal Trellis Construction of Linear Block Codes.

In this part, we present two methods for constructing the minimal trellis of linear block codes. We first describe the BCJR algorithm, followed by the Massey algorithm. Then, the minimal trellis construction of systematic-like RM codes will be discussed.

6.3.2.1 BCJR Construction. The idea behind the BCJR trellis construction of block codes is based on the parity check matrix and syndrome calculation. It uses the fact that the row space of the parity check matrix is in null space of the rows of generator matrix.

Let $(n-k) \times n$ parity check matrix, H, be as follows:

$$H = \begin{bmatrix} h_0 & h_1 & \cdots & h_{n-1} \end{bmatrix} \qquad (6.9)$$

where h_i , $i = 0, 1, \ldots, n-1$ are column vectors of the parity check matrix H

In a linear block code, a valid codeword, $C = (c_0, c_1, \cdots, c_{n-1})$, $C \in \mathbf{C}$ has to satisfy the following constraint, i.e. syndrome, $S = \mathbf{0}$.

$$S = CH^T = c_0 h_0 + c_1 h_1 + \cdots + c_{n-1} h_{n-1} = \mathbf{0} \qquad (6.10)$$

where $\mathbf{0}$ represents the $(n-k)$ zero-column vector. By following the above relationship, the trellis of a linear block code can be constructed.

Let $s' \in S_i$ and $s \in S_{i+1}$ be the states of the trellis at depth i and $i+1$, respectively. We have,

$$s = s' + \alpha_i h_i \qquad (6.11)$$

where $\alpha_i \in \{0, 1\}$, $i = 0, 1, ..., n-1$ for binary linear code. Generalization for non binary linear code is straightforward. Under the condition in Equation (6.10), the final state must satisfy the following equation,

$$s_f = \sum_{i=0}^{n-1} \alpha_i h_i = \mathbf{0} \tag{6.12}$$

As an example, consider the (7,4) Hamming code with the parity check matrix given by Equation (6.13). The trellis diagram for this code is shown in Figure 6.1. The codeword $C = (1,0,1,1,0,1,0)$ is represented by the labels of the dashed path starting from the zero state and ending at the zero state.

$$H = [\ h_0 \quad h_1 \quad \cdots \quad h_{n-1}\ ] = \begin{bmatrix} 0 & 1 & 1 & 1 & 1 & 0 & 0 \\ 1 & 0 & 1 & 1 & 0 & 1 & 0 \\ 1 & 1 & 0 & 1 & 0 & 0 & 1 \end{bmatrix} \tag{6.13}$$

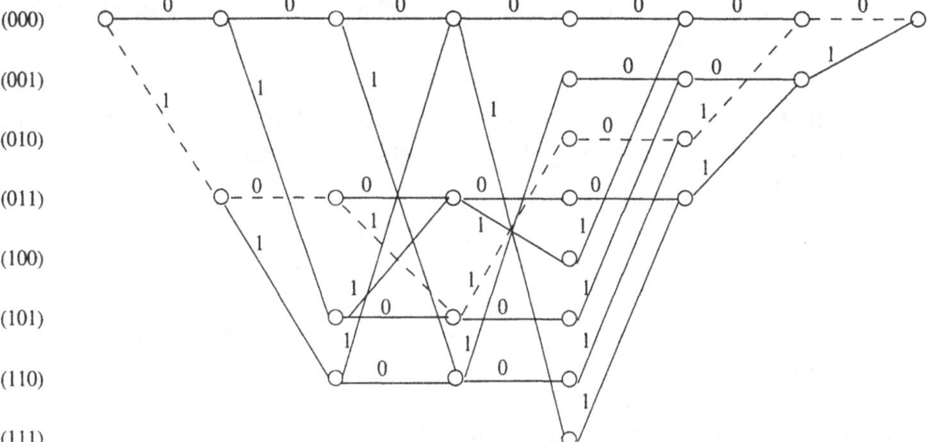

Figure 6.1. Trellis Diagram of the (7,4) Hamming Code

6.3.2.2 Massey Construction. A code C is constructed from a $(k \times n)$ generator matrix G of reduced echelon form. Let $\gamma_j \ : \ j = 0, 1, ..., k-1$ be the left indices of matrix G. As it was stated in the properties of a reduced echelon matrix, the k columns at the left indices have weight one. This implies that in a codeword, the information bits can be found at positions of the left indices.

Trellis $T = (S, B, L)$ of the block code, $\mathbf{C}$ over $GF(q)$ is constructed by specifying the set of states S_i at time $i \ : \ i = 0, 1, ..., n$. The states in S_i are identified by the knowledge of the information symbols already observed at time i, thus, all other information symbols are assumed to be zero. Let p be the largest index such that $\gamma_p \leq i$. States in S_i are labeled by

$$s_j = \{(c_{i+1}, ..., c_n) \ : \ (c_1, ..., c_n) = (m_1, ..., m_p, 0, ..., 0)G\} \tag{6.14}$$

where $m_1, ..., m_p$ are the first p information bits. The original and the final states are $S_o = \{0\}$ and $S_n = \{\phi\}$ by tradition, where ϕ is an empty string.

The branches in B_i of T are defined as follows :

- When $i = \gamma_p$, there is a branch $b \in B_i$ connecting a state, $s \in S_{i-1}$ with a state, $s' \in S_i$, if and only if there exist codewords $c = (c_1, c_2, ..., c_n)$ and $c' = (c'_1, c'_2, ..., c'_n)$ in C such that,

$$\begin{aligned} (c_i, c_{i+1}, ..., c_n) &= s \\ (c'_{i+1}, ...c'_n) &= s' \end{aligned} \tag{6.15}$$

  and either $c' = c$, or $\beta(c' - c)$ is equal to the pth row of G for some $\beta \in GF(q)$. The branch label is c'_i and the number of out-going branches at each state in S_{i-1} is q.

- When $i > \gamma_p$, there is a branch $b \in B_i$ connecting a state $s \in S_{i-1}$ with a state $s' \in S_i$ if and only if there exists a codeword $(c_1, c_2, ..., c_n) \in C$ such that,

$$\begin{aligned} (c_i, c_{i+1}, ..., c_n) &= s \\ (c_{i+1}, ..., c_n) &= s' \end{aligned} \tag{6.16}$$

  The branch label is c_i . In this case there is only one out-going branch from each state in S_{i-1}.

6.3.2.3 Trellis Diagram of the RM Code. The generator matrix of the RM code is constructed as defined in Section 6.2 and modified to be in a row-reduced echelon form. After that, the Massey algorithm is applied to the matrix to construct the minimal trellis of the code. As stated in the previous part, the positions of the information bits can be indicated in a codeword even though they are not at first or last k positions of a codeword as in a systematic code. Thus, we can consider the code as a systematic-like RM code. Following is the generator matrix, G of an RM (8,4) code and a trellis diagram of the code drawn by Massey algorithm is shown in Figure 6.2. The example of a codeword $C = (1, 1, 0, 0, 0, 0, 1, 1)$ is represented by a sequence of branch labels of the path with dashed lines as shown in the same figure.

$$G = \begin{bmatrix} 1 & 0 & 0 & 1 & 0 & 1 & 1 & 0 \\ 0 & 1 & 0 & 1 & 0 & 1 & 0 & 1 \\ 0 & 0 & 1 & 1 & 0 & 0 & 1 & 1 \\ 0 & 0 & 0 & 0 & 1 & 1 & 1 & 1 \end{bmatrix} \tag{6.17}$$

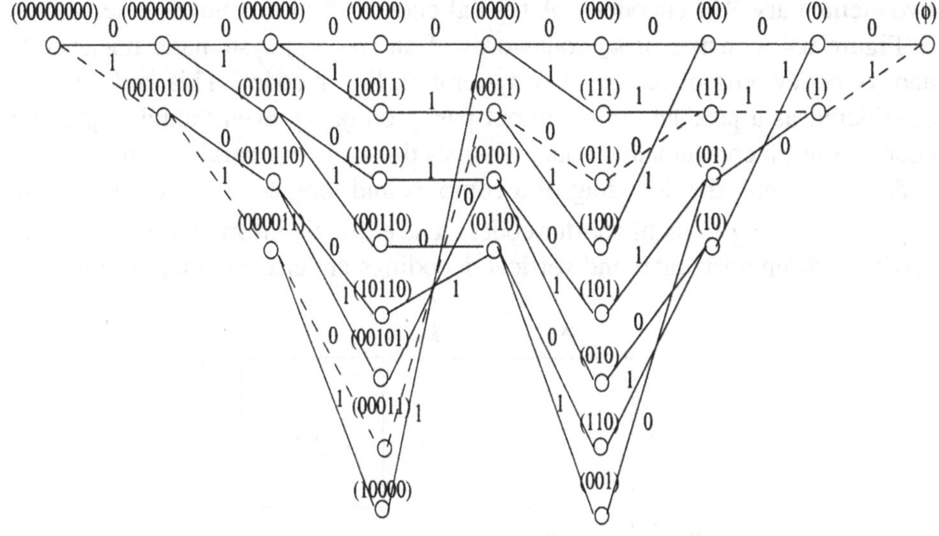

Figure 6.2. Trellis Diagram of the RM (8,4) Code

6.4. Reed-Muller Turbo Codes

In this section, *Reed-Muller* (RM)-turbo codes are discussed. As mentioned earlier, the minimal trellis of RM codes are known. Moreover, the Massey algorithm can be used to construct the trellis diagram of systematic-like RM codes. These considerations make RM-turbo codes suitable for use with trellis-based iterative decoding. We present the details of RM-turbo codes and their applications in satellite ATM transmission.

6.4.1 RM Turbo Encoder

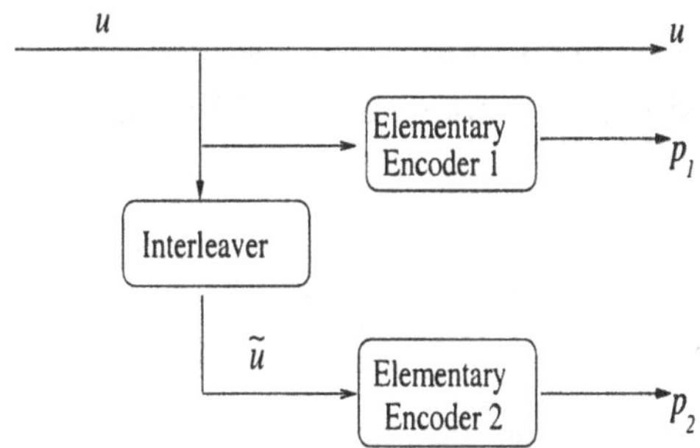

Figure 6.3. RM-turbo Encoder

In this book, we consider a parallel two-dimensional block turbo encoder shown in Figure 6.3. Information block u and its permuted version $\tilde{u}$ are encoded by two elementary RM encoders. A typical codeword of this turbo code, shown in Figure 6.4, consists of k_2 codewords of an (n_1, k_1) systematic linear code and k_1 codewords of an (n_2, k_2) systematic linear code. This code can be considered as a product code without parity on parity even though a product code is a serial concatenated code, whereas this code is a parallel concatenated code. Therefore, the decoding of a complex and long code can be broken up into the decoding steps of shorter codes. The extrinsic information L_e^{-} and $L_e^{|}$ produced from horizontal and vertical decodings are also shown in Figure 6.4.

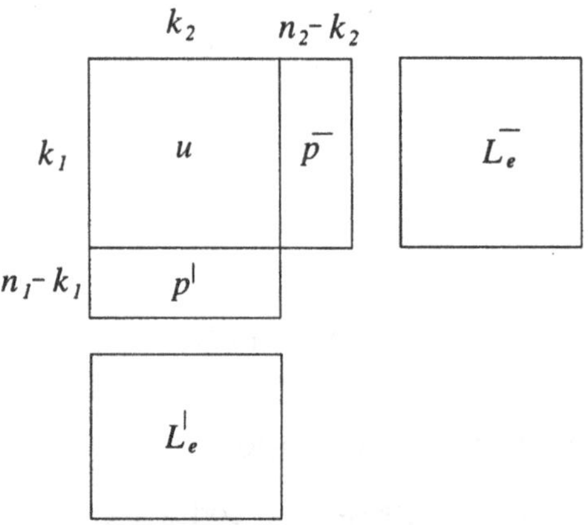

Figure 6.4. Two-dimensional Block Code

The $k_1 \times k_2$ bits long information block, u, is encoded horizontally using the (n_2, k_2) elementary encoder 1 generating k_1 codewords by adding the parity block p^{-}. Then, the u is encoded vertically. This can be thought of as the interleaved version of horizontal one resulted from block interleaving. This operation generates k_2 codewords of an (n_1, k_1) code by adding the parity block $p^{|}$.

We consider the two-dimensional RM code, which uses the same RM (n, k) code in each dimension and denote it as an RM $(n, k)^2$ code. The overall code rate is given as,

$$R = \frac{k^2}{n^2 - (n - k)^2} \tag{6.18}$$

Figure 6.4 demonstrates the idea of a two-dimensional codes in the case of systematic component codes. We use RM codes in which the information part, although distinct, is not placed at the beginning of the codeword. Figure 6.5 shows a systematic-like RM code. The shaded areas are information regions. Only the systematic-like structure of the horizontal code is shown in this figure. For presentation convenience, some modification on horizontal code has been made by reordering code bits to obtain the sequence whose first k positions are information bits. Along similar lines, the vertical code is also reordered. Note that the reordered sequences are not codewords, thus the inverse operation has to be performed before the decoding process starts. Figure 6.4 can represents the reordered sequences.

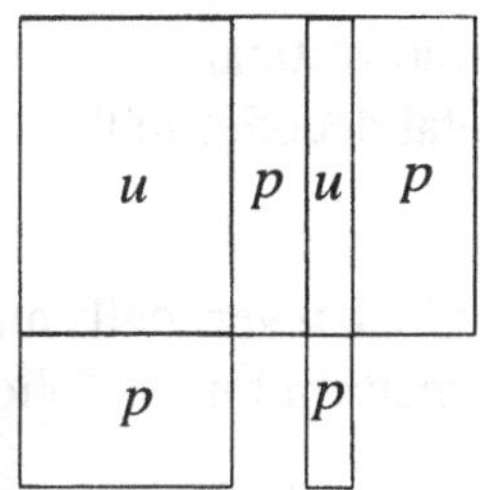

Figure 6.5. Systematic-like RM Code

6.4.2 Turbo Decoder

In this section, the iterative decoding for two-dimensional block codes is discussed. The SISO decoders use the trellis-based MAP decoding algorithm presented in Section 5.3.4.

6.4.3 Iterative Decoding of a Two-Dimensional Code

Figure 6.6 shows the iterative decoding procedure of the code shown in Figure 6.4.

1 Set the a priori value $L(u) = 0$. Set the iteration number $I = 0$.

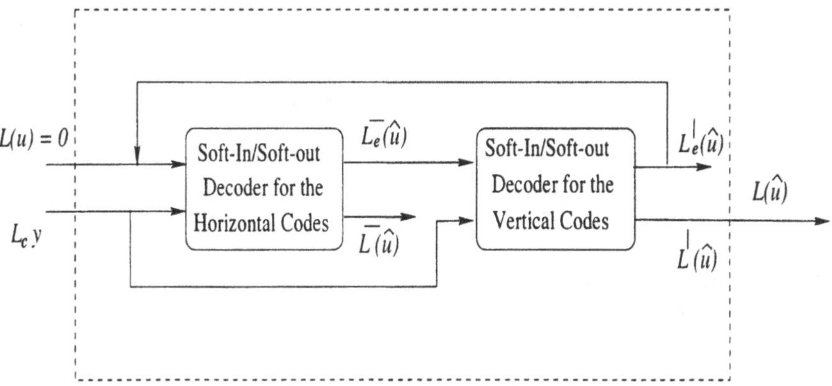

Figure 6.6. Iterative Decoding Procedure of Two-dimensional Block Code

2 Decode the information block u horizontally and obtain the horizontal extrinsic information for the information bits as follows:

$$L_e^-(\hat{u}) = L^-(\hat{u}) - L_c \cdot y - L(u) \tag{6.19}$$

3 Set $L(u) = L_e^-(\hat{u})$ for the vertical decoder, i.e., pass the extrinsic information from the horizontal decoder to the vertical decoder as the a priori value of information bits.

4 Decode the information block u vertically and obtain the vertical extrinsic information for the information bits as follows:

$$L_e^|(\hat{u}) = L^|(\hat{u}) - L_c \cdot y - L(u) \tag{6.20}$$

5 Set $L(u) = L_e^|(\hat{u})$, i.e., pass the extrinsic information from the vertical decoder to the horizontal decoder as the a pririori value.

6 If $I < K$, set $I = I + 1$ otherwise go to step 7.

7 The soft output is:

$$L(\hat{u}) = L_c \cdot y + L_e^-(\hat{u}) + L_e^|(\hat{u}) \tag{6.21}$$

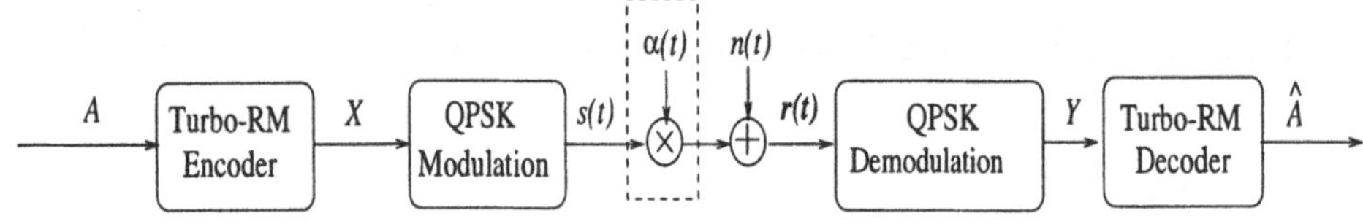

Figure 6.7. System Model

6.4.4 System Model

Figure 6.7 shows the channel model used for the simulation purpose. The received signal can be written as

$$r(t) = \alpha(t)s_i(t) + n(t) \tag{6.22}$$

where $\alpha(t)$ is a Rayleigh process, i.e., at any specific time it is a random variable α that satisfies $E(\alpha^2) = 1$, and has the probability density function,

$$p_\alpha(\alpha) = 2\alpha e^{-\alpha^2} \quad , \; a \geq 0 \tag{6.23}$$

and $n(t)$ is a white Gaussian noise process with two-sided power spectral density $N_o/2$. We assume that QPSK modulation is used. $s_i(t) \; : i = 1, 2, 3, 4$ is the modulated waveform for the symbol s_i . We consider the following cases:

AWGN channel :	$\alpha(t) = 1$
Fading channel:	$\alpha(t)$ is Rayleigh process

6.4.5 Simulation Results

Figure 6.8 shows the BER versus E_b/N_o of the RM $(32, 26)^2$-turbo code for different number of iterations. The coding gain obtained by increasing the number of iterations is high at first, but, saturates later and after the 5th iteration the performance improvement for RM $(32, 26)^2$ is negligible.

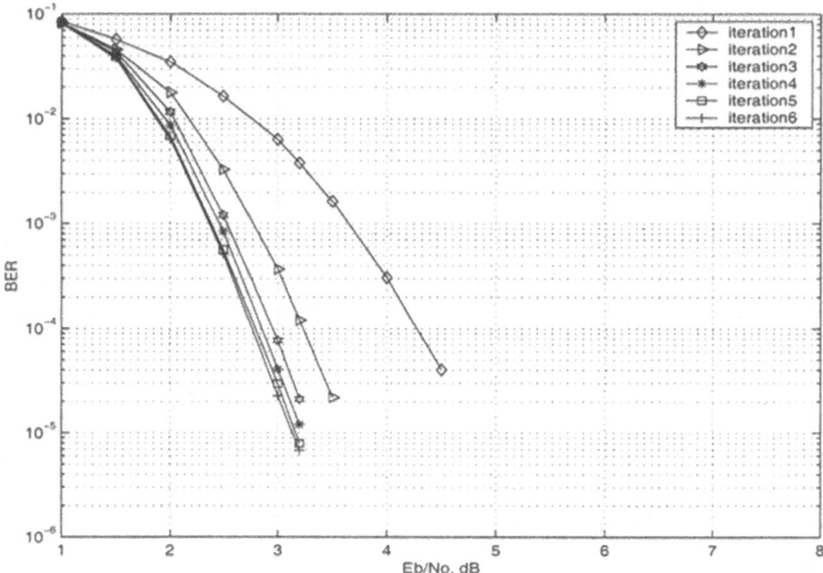

Figure 6.8. Performance of a RM $(32, 26)^2$-turbo Code with Different Iterations on an AWGN Channel

In [159], it is shown that the performance saturation occurs sooner for shorter turbo codes. For example, the saturation for RM $(8, 4)^2$, RM $(16, 11)^2$ and RM $(32, 26)^2$ codes occur after 2,4 and 5 iterations, respectively. The reason for faster saturation of the performance of the turbo codes with shorter component codes is that a shorter code has a smaller size interleaver, i.e., less information diversity. In the rest of this chapter, the number of iterations used in the simulations of RM $(8,4)^2$, RM $(16,11)^2$ and RM $(32,26)^2$-turbo codes are two, four and five, respectively.

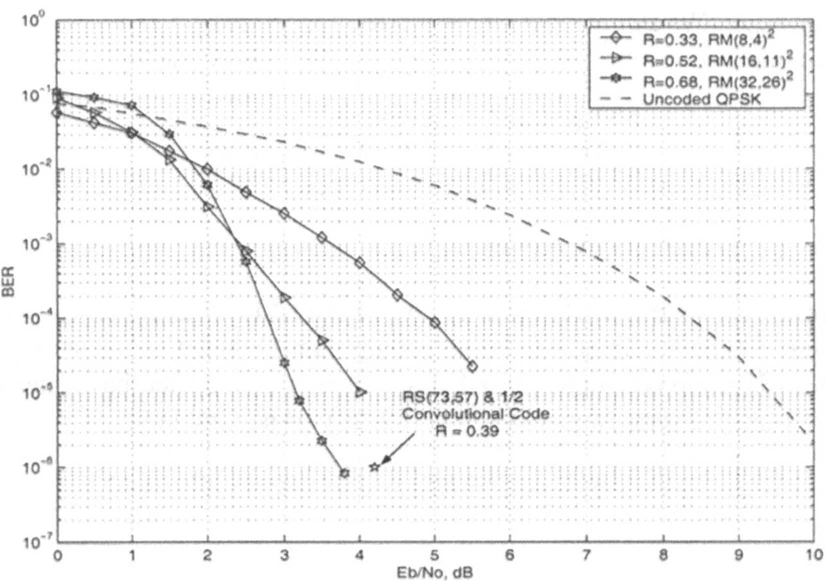

Figure 6.9. Performance of RM-turbo Codes with Different Code Lengths after 5 Iterations on an AWGN Channel

The BER versus E_b/N_o curves of RM-turbo codes with different lengths in an AWGN channel are shown in Figure 6.9. The coding gains of 3.6. 5.3 and 6.2 dB for RM $(8,4)^2$, $(16,11)^2$, $(32,26)^2$ turbo codes are obtained over the uncoded QPSK at BER of 10^{-5}. The results are better than the ones reported in [7] as expected since we are using stronger component codes. RM codes used can be considered as extended Hamming codes as opposed to the Hamming codes used in [7]. In addition, the performance of RS(73,57) and convolutional code rate of $\frac{1}{2}$ concatenated code at BER of 10^{-6} is given and compared with RM $(32,26)^2$ turbo code where the turbo coding scheme obtains coding gain of about 0.5 dB with higher code rate of about 1.7 times that of the concatenated code, which is equivalent to an additional coding gain of about 2.4 dB [1].

Figure 6.10 depicts the BER versus E_b/N_o of RM-turbo codes in Rayleigh fading channel. We obtain at least 25 dB coding gain over the uncoded QPSK for RM $(8,4)^2$, $(16,11)^2$, $(32,26)^2$- turbo codes at BER of 10^{-4}. It is shown

[1]Assuming that power and bandwidth are equally valuable

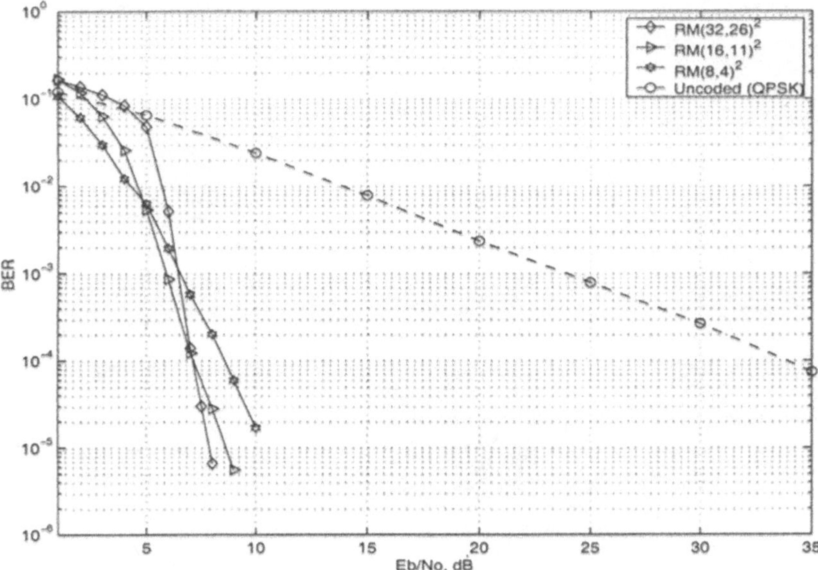

Figure 6.10. Performance of RM-turbo Codes with Different Code Lengths after 5 Iterations on a Rayleigh Fading Channel

in Figure 6.11 that a higher coding gain is obtained for more iterations and saturation occurs after 5 iterations in the case of RM $(32, 26)^2$-turbo code.

6.4.6 Design of RM Turbo Codes for Satellite ATM

The basic format of an ATM cell is shown in Figure 6.12. The ATM cell consists of 53 bytes which is divided into two parts. The first 5-byte part is the cell-header, and the 48-byte part is the payload or the user information. For satellite applications, an ATM traffic cell contains an extra 4-byte header part which is the request sub-field of the Satellite Access Control (SAC) [28].

6.4.6.1 Shortening Patterns for the RM Turbo Codes. The ordinary shortening pattern of a given code is to set some of the bits at the end of the message equal to zero. However, it is not mandatory, so in this part we propose different patterns for shortening of the RM-turbo codes.

In our study, we consider ATM cells for satellite applications. These cells have 48-byte payload and 9-byte header. Thus, we use a two dimensional code with RM (32,26) component codes. The information section of the resulting turbo code is $26^2 = 676$ bits long which exceeds the length of a satellite ATM cell, i.e., $57 \times 8 = 456$ bits by 220 bits. That is, we have to shorten the RM-turbo code by setting 220 information bits to zero. These zeros will not be transmitted, hence, the overall rate of the code, R_c will roughly be,

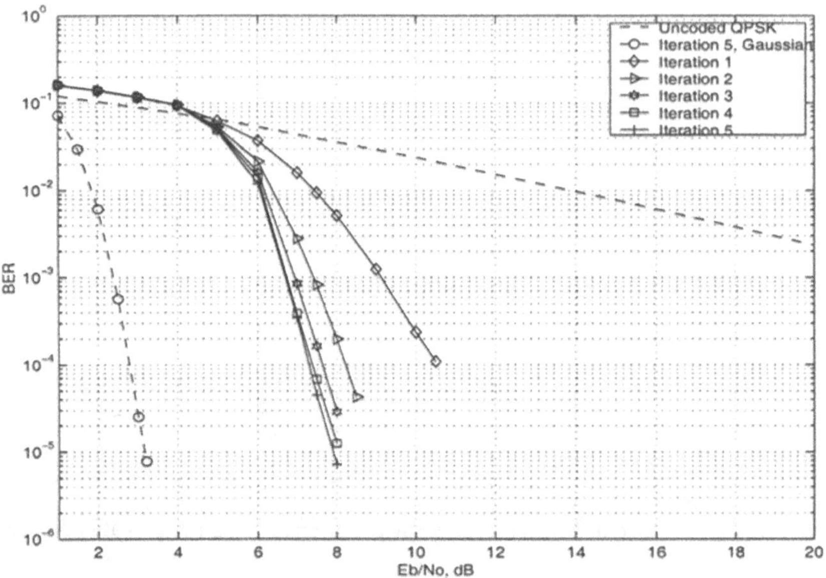

Figure 6.11. Performance of RM $(32, 26)^2$-turbo Code with Different Number of Iterations on a Rayleigh Fading Channel

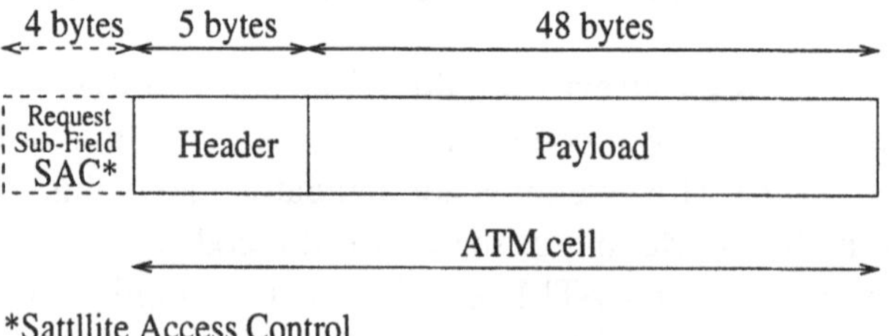

Figure 6.12. Satellite ATM Cell

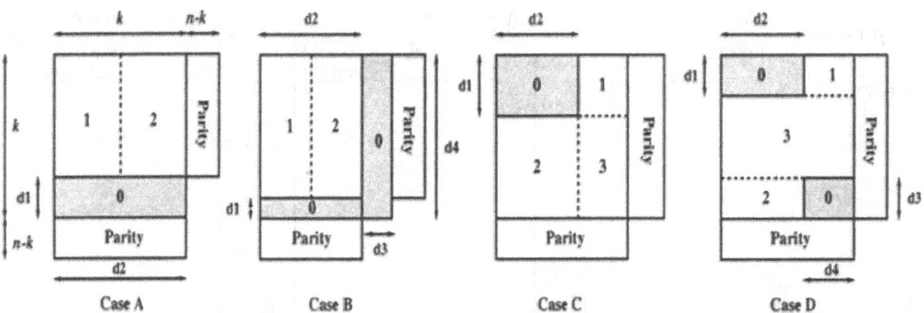

Figure 6.13. Shortening Patterns

$$R_c = \frac{456}{32^2 - 6^2 - 220} \cong 0.594 \tag{6.24}$$

Of course, the exact rate depends on the particular shortening scheme. Figure 6.13 shows four shortening patterns. The following are the different shortening patterns with their corresponding rates :

Case A : d1 = 8 , d2 = 26, $R_c = 0.64$
Case B : d1 = 2, d2 = 19, and d3 = 7 , d4 = 26 ,$R_c = 0.64$
Case C : d1 = 11 , d2 = 18 , $R_c = 0.61$
Case D : d1 = 8 , d2 = 17 and d3 = 8 , d4 = 9 , $R_c = 0.60$

where d1, d2, d3 and d4 are the dimensions of the deleted blocks shown in Figure 6.13.

In cases A and B, we attempt to design the shortening patterns in such a way that we can reduce the number of parity bits to obtain higher code rate. However, in cases C and D, we design the shortening pattern to construct codes having the special property of UEP by surrounding the highly protected part with more zeros than others as shown in region 1 of the cases C and D.

6.4.6.2 Simulation Results. Figure 6.14 and 6.15 show the performance at different regions in the shortened codes. The shortening pattern in case A is an ordinary shortening pattern, whereas cases B, C and D are modified shortening patterns. In cases A and B, the performance of different regions is almost the same. This shows the *Equal Error Protection* (EEP) property of these codes. In contrast, the performance of cases C and D in different regions is different depending on the number of zeros (shortened bits) surrounding the region. The lower BER in the regions that have more zeros around them is observed. The UEP property is obviously seen in cases C and D where the region 1 gets the best performance followed by regions 2 and 3.

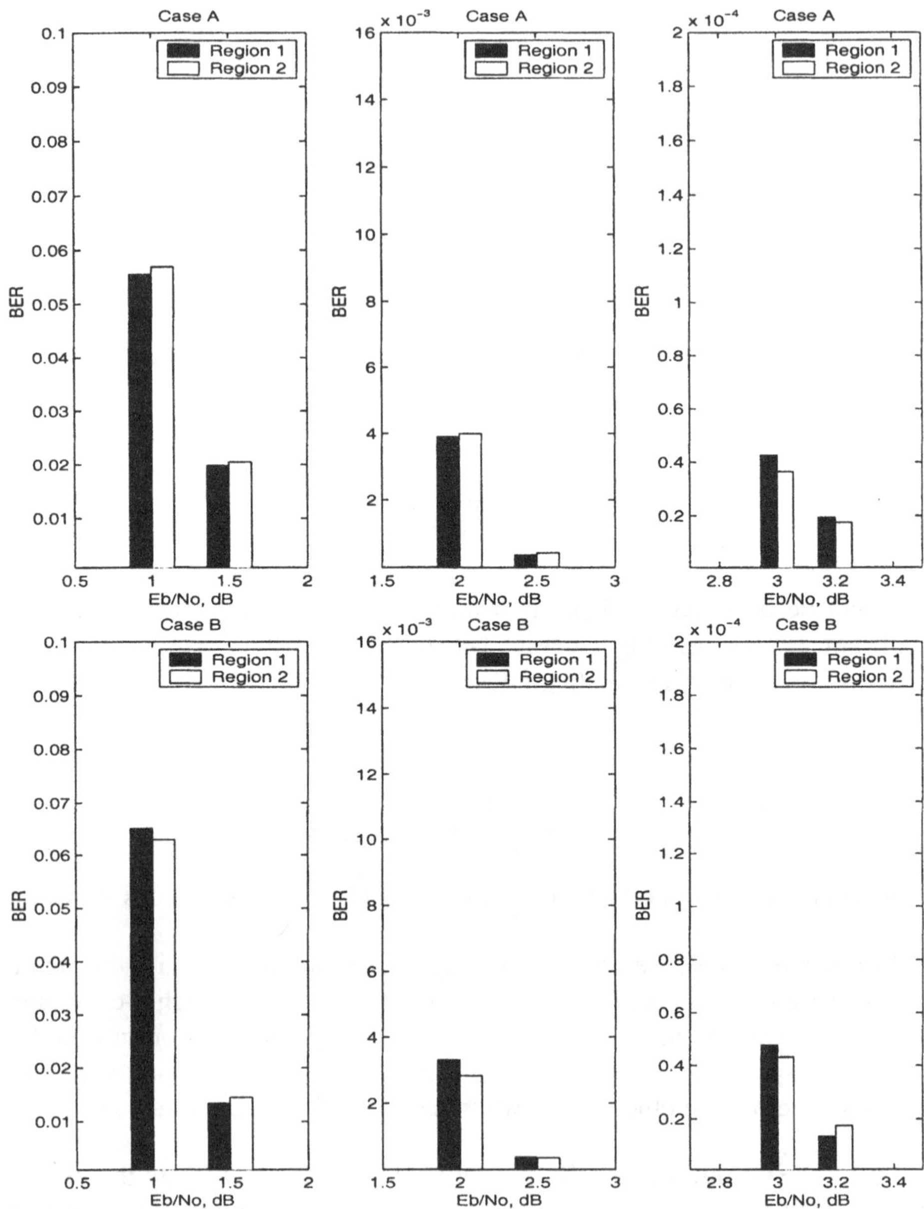

Figure 6.14. Performance of Shortening Patterns A and B at Different Regions.

Figure 6.16 compares the BER of two different regions in case C with the performance of case B. The results show that the BER of the best region in UEP code is lower than that of EEP code. Figure 6.17 shows the overall performance of the EEP, UEP and the original (un-shortened) codes, where the EEP codes provide a coding gain of about 0.2 dB over UEP codes. Also, the performance

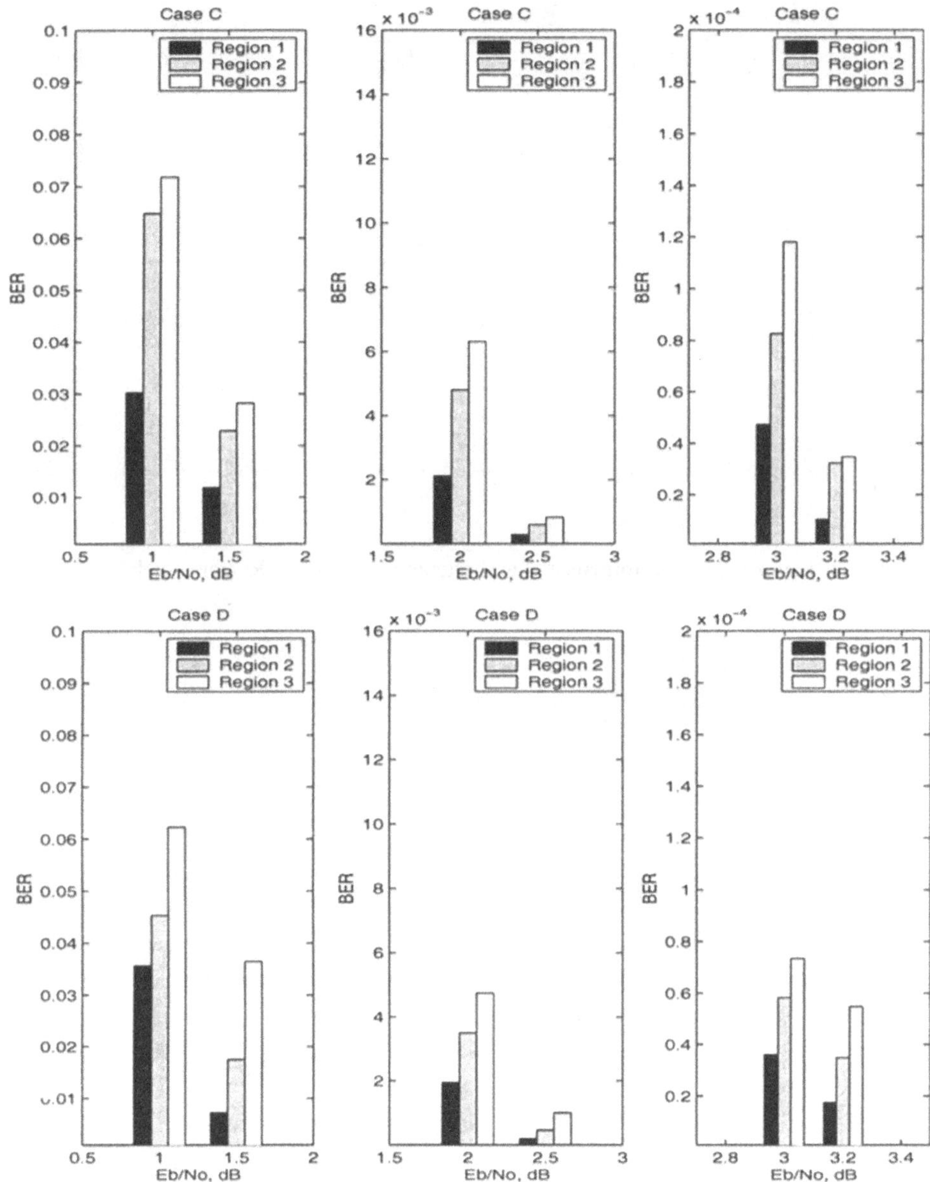

Figure 6.15. Performance of Shortening Patterns C and D at Different Regions.

of the EEP codes and UEP codes are about 0.2 and 0.4 dB worse than that of the original RM-turbo codes. This is so, because the shortening process affects the distance spectrum of the two-dimensional codes.

The performance comparison among different coding schemes for ATM transmission is illustrated in Figure 6.18. The performance of the proposed

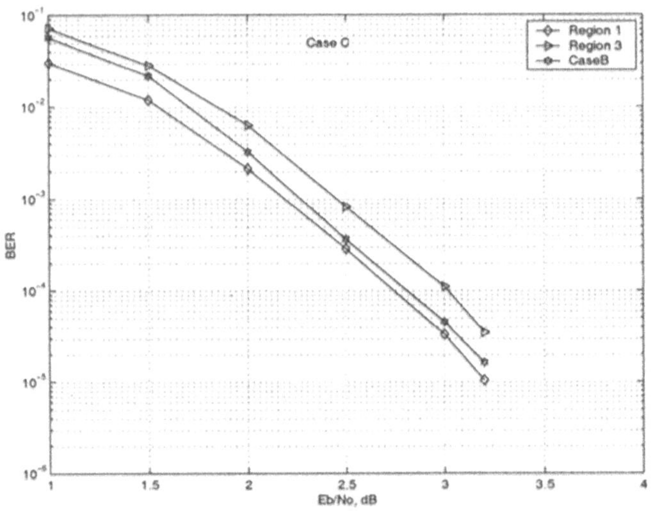

Figure 6.16. Performance of a Shortening Pattern B at Region 1 and 3

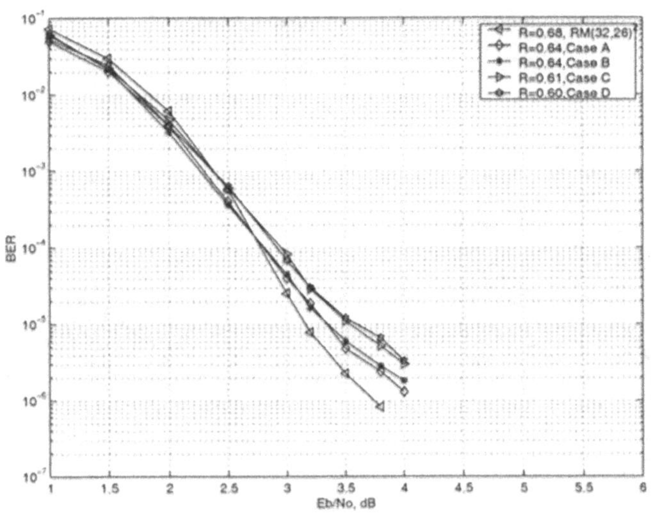

Figure 6.17. Overall Performance of Shortened RM-turbo Codes with Different Shortening Patterns

coding scheme case A is compared with the conventional scheme, RS(73,57) and convolutional rate of $\frac{1}{2}$ concatenated code [148]. It is shown that the coding gain of about 0.2 dB is obtained by the proposed coding scheme over the concatenated code at BER of 10^{-6} and the code rate of the proposed code is higher than that of the concatenated code by about 1.6 times which is equivalent to an additional coding gain of about 2 dB. Furthermore, the lower bound of BER of the double-binary Circular Recursive Systematic Convolutional (CRSC) turbo code after 8 iterations with 4-bit quantization is given. This lower bound of BER is calculated from the block error rate presented in [36]. The CRSC code obtains a coding gain of about 0.8 dB over the RM-turbo code case A. Finally, the performance of the shortened version of the extended Hamming $(32, 26)^2$ code at BER of 10^{-5} is given. This coding scheme performs worse than the proposed coding scheme by about 0.3 dB with a lower code rate that makes it equivalent to an overall degradation of about 0.5 dB.

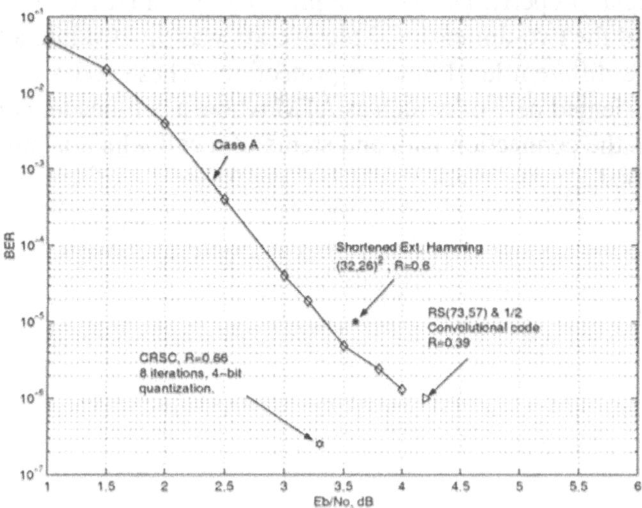

Figure 6.18. Performance Comparison of Different Coding Schemes for ATM Transmission

6.5. Summary

In this chapter, first the definitions and the code construction of RM codes were discussed. Then, the minimal trellis constructions using BCJR and Massey algorithms were presented. The minimal trellis is the best trellis representation of a block code in the sense of having the lowest number of states at each time index. The example of the construction of RM(8,4) code and its trellis representation were given. Then, the RM turbo encoder was described which could be considered as a product code without parity on parity. The information was encoded both horizontally and vertically, where the vertical version of the information could be considered as the permuted version of information with block interleaver. In the decoding process, iterative MAP decoding of the two-dimensional block code was discussed. The simulation results on AWGN and Rayleigh-fading channels were presented. It was shown that the turbo decoding improves the performance when the number of iterations increases, although it saturates after a few iterations. The number of iterations needed depends on the interleaver size. The longer the interleaver is the more gain is obtained from increasing the number of iterations. The saturation is a result of the extrinsic information exchanged between the two decoders being highly correlated so that no extra information could be provided after a few iterations.

We also presented RM turbo codes for satellite ATM applications. The shortened RM turbo codes with four different shortening patterns were discussed. Two of the shortening patterns were designed to reduce the number of parity bits of the codes resulting in higher rate codes. The other two were designed to obtain special property, i.e. the UEP property. In an UEP code, the information portion of the two-dimensional block code was divided into a few regions, each having a different level of error protection. These codes are suitable for connection-oriented networks such as ATM where a cell-header contains information about the connection path and status of a cell which is more important than its payload.

Chapter 7

PERFORMANCE OF BTCs AND THEIR APPLICATIONS

7.1. Introduction

In this chapter, we present some results of BTCs from the literatures and the applications of BTCs or TPCs, in particular. Performance of BTCs using trellis-based and algebraic-based decoding methods are presented. Different constituent codes for BTCs, e.g. BCH, Hamming and single parity check codes are used. BTCs show excellent performance as TCCs do. Thus, the actual power is in the iterative decoding algorithm rather than the codes. BTCs have the benefits of being simple, performing well in high code rate systems and showing less error floor effects due to their large minimum distance. Some results are investigated for AWGN channel, flat Rician and flat Rayleigh fading channels which mimic the real satellite and wireless communication links. In the next generation communication systems, coding schemes sought are not only for operating at low SNR but also for providing high spectral efficiency. Therefore, the BTCs with different modulation schemes are presented. Moreover, different information lengths, i.e. interleaver size of BTCs are investigated. Comparison of TCCs and TPCs are also shown. Finally, the applications of BTCs, focusing on the wireless and satellite communication are presented.

7.2. Some Results from the Literatures

Simulation results of BTCs using BCH and RS codes as constituent codes have been presented in Section 5.3.3.6. The decoding algorithm used is based on modified Chase algorithm. More results from the literatures will be presented in this section.

Performance of Hamming-BTCs is shown in Figure 7.1. This result was presented in [7]. The trellis-based iterative decoding algorithm is used. It is also noted that the parallel concatenated turbo codes and trellis-based decoding algorithm presented in Sections 5.2.2 and 5.3.4 follow this work. Six iterations are run. It is observed that the longer the constituent codes, the better the performance in high SNR region. The reason is that a longer code has a longer interleaver. Furthermore, the results also show that the BTCs with high code

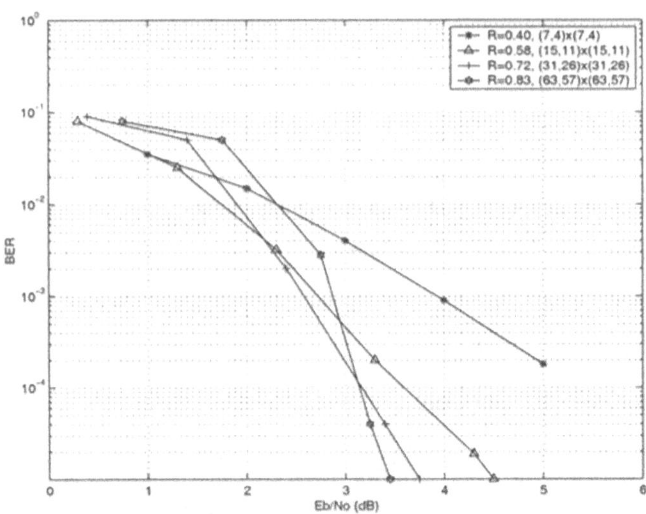

Figure 7.1. Performance of Hamming-BTCs over AWGN Channel

rate perform well. This confirms the claim in [128],[10] that BTCs are good for high code rate applications.

Performance comparison of TCC and TPCs under different channel conditions and different modulation schemes are presented in [149] and reproduced in this book. The results are illustrated in Figure 7.2-7.4. System configuration is given as follows. In the case of TCCs, the codes and the interleaver used are as specified in UMTS standard and are decoded with the max-log MAP decoding algorithm. In the case of TPCs or serial concatenated block codes, BCH codes are used as constituent codes with the decoding algorithm presented in [124],[125]. The code rate is $\frac{1}{2}$. In each system, two information block sizes, viz., 424 and 848 bits, for one and two ATM cells, are investigated. TPCs based on $(26, 20) \times (31, 20)$ and $(42, 29) \times (42, 29)$ shortened BCH codes are used for one and two ATM cells, respectively. Figure 7.2 illustrates the performance comparison of TCCs and TPCs using QPSK modulation for AWGN channel. It is noticed that similar performance for TPCs and TCCs is obtained in the case of one ATM cell. However, TCC performs better than TPC when block size increases in the low and moderate SNR region before the effect of error floor is observed and TPC will take over TCC in high SNR region.

Performance comparison in the case of 16-QAM modulation over AWGN channel is shown in Figure 7.3. TCCs outperform TPCs in both cases, with coding gains of about 0.75 dB and 0.5 dB for one and two ATM cells, respectively.

The performance of the codes over flat Rician (K=5) and Rayleigh fading channels is presented in Figure 7.4. The former channel models the communi-

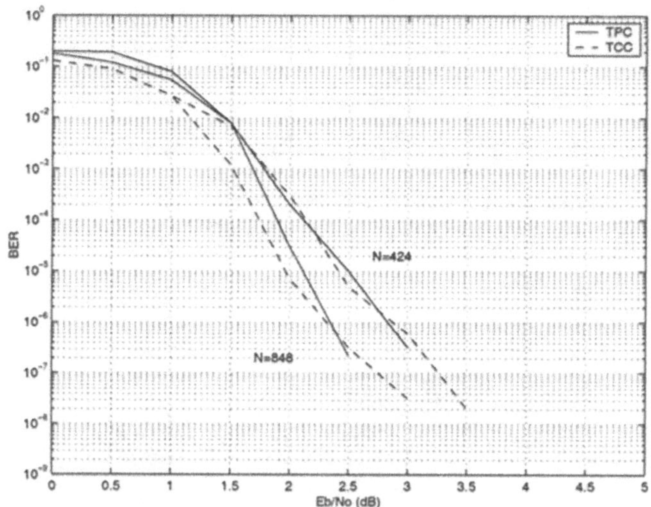

Figure 7.2. Performance Comparison of TCC and BTC for QPSK, AWGN Channel, Rate 1/2

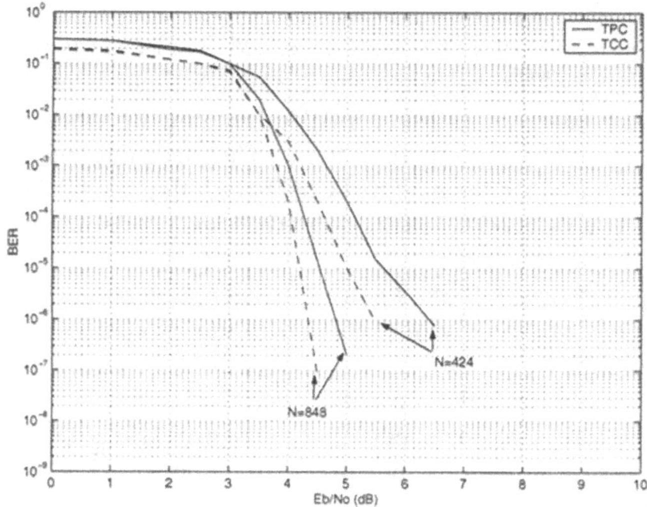

Figure 7.3. Performance Comparison of TCC and BTC for 16-QAM, AWGN Channel, Rate 1/2

cation link between mobile terminal and the satellite, whereas the latter channel models the link between mobile terminal and the base-station. AWGN channel models the communication between fixed terminal and the satellite. TPCs outperform TCCs in flat fading Rayleigh channel because of the error floor of TCCs at high SNR. TCCs and TPCs perform similarly in the case of flat fading Rician channel.

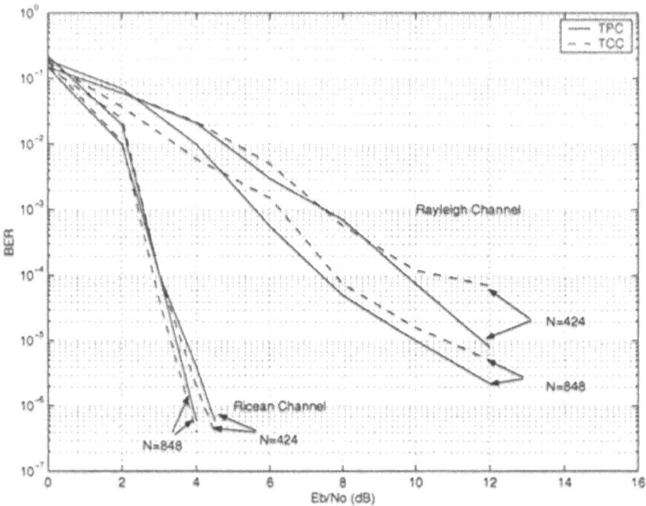

Figure 7.4. Performance Comparison of TCC and BTC for QPSK, Fading channels, Rate 1/2

Simulation results of TPCs using extended BCH code with a code length of 32 bits along with their Maximum Likelihood bounds presented in [126] are shown in Figure 7.5. In [126], FBBA decoding algorithm, an *Augmented List Decoding* (ALD) algorithm, is used. This decoding algorithm provides better performance than the modified Chase algorithm used in [10] with higher complexity. The BCH-TPC $(32, 21)^2$ or (1024,441) code at rate of 0.43 can be used in ATM applications.

A more general code design known as *Generalization Turbo Product Code* (GTPC) is presented briefly in [126] and in more detail in [127]. The idea is to use more than one row constituent code in the product code, where one column code is used as in the conventional product code. The choice of the row codes need to satisfying the condition discussed in [127]. By using this code construction, UEP property is obtained, i.e., stronger codes are used for the highly protected information part. As it was mentioned earlier, one application of UEP codes is in ATM transmission. The performance of the UEP code is illustrated in Figure7.6, where lower BER curves are for header and the higher BER curves are for payload.

7.3. Applications of Block Turbo Codes.

In this section, the applications of BTCs, mainly in satellite communications, wireless *Local Area Network* (LAN), wireless internet access and mobile communication are presented. In this section, we introduce the organizations and companies that propose BTCs as the standard or use them in their products.

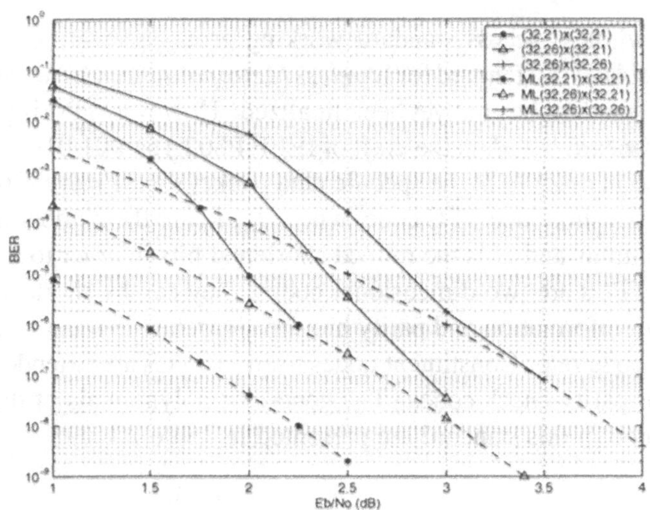

Figure 7.5. TPC with Extended BCH of n=32

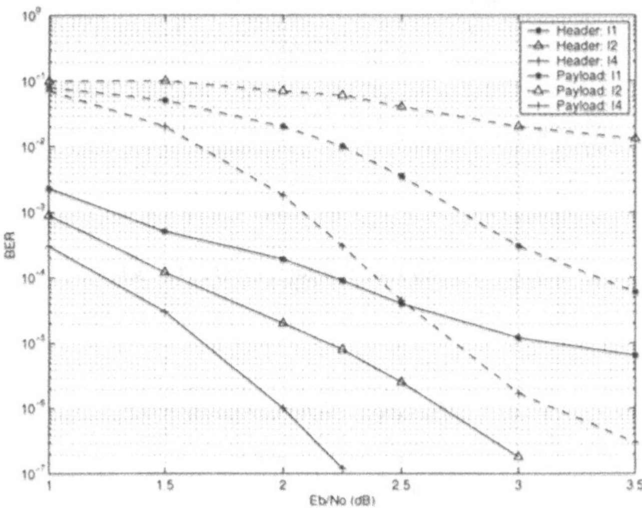

Figure 7.6. GTPC with UEP Performance

BTCs that so far have been implemented are product codes with iterative decoding algorithm based on different variations of the algebraic decoding algorithm or modified Chase-II algorithm rather than the trellis-based algorithm. This can be justified by the lower complexity of these algorithms and their near optimal performance at moderate to high SNRs.

7.3.1 Broadband Wireless Access Standard

Broadband wireless access standard working group is creating IEEE 802.16 $WirelessMAN^{TM}$ Standard for *Wireless Metropolitan Area Networks (MAN)*. To date, the standard has not yet been finalized. However, TPC is proposed for use in this standard based on the draft of IEEE P802.16ab-01/01r2 [153]. TPCs are shortened in order to fit the specific data packet size. Figure 5.14 shows the structure of the shortened TPC code, where I_x, I_y are the number of shortened rows and columns and B is the refined shortened bits in order to match to data size. Since TPCs are two dimensional codes, they can be shortened in such a way that both information and parity bits are shortened in order to obtain highest possible code rate. Constituent codes used in TPCs are extended Hamming codes and/or parity check codes. The generator polynomials of the Hamming codes specified in this standard are presented in Table 7.1.

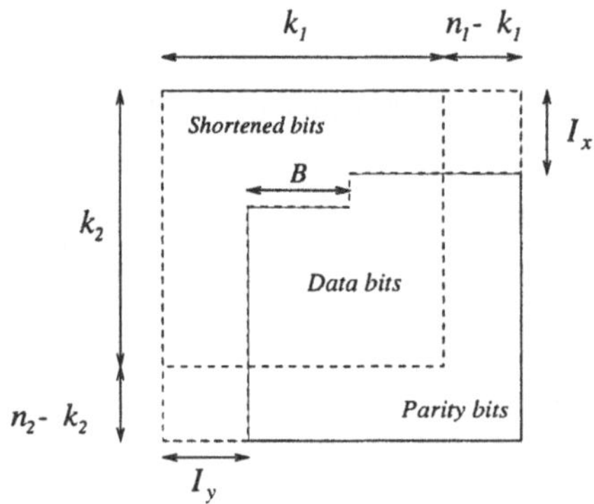

Figure 7.7. Structure of Shortened 2D Block

Code Length	*Message Length*	*Generator polynomial*
7	4	$X^3 + X + 1$
15	11	$X^4 + X + 1$
31	26	$X^5 + X^2 + 1$
63	57	$X^6 + X + 1$

Table 7.1. Hamming Code Generator Polynomials

Three types of interleavers are used [153]:

- Type 1 (no interleaver): Bits are written row-by-row and read row-by-row. Therefore, there is no additional latency.
- Type 2 (block interleaver): Encoded bits are read from the encoder only after all first k_2 rows are written into the encoder memory. The bits are read column-by-column from top to bottom. There is one block of additional latency using this interlever.
- Type 3 (permutation interleaver): Reserved. There is possibility of including helical interleaver.

The recommended TPC codes for this standard are presented in Table 7.2 Decoding algorithms used are not specified and one may use any available SISO algorithm. However, the standard states three algorithms, viz., SOVA, a soft-output variant of the modified Chase algorithm and the BCJR algorithm

Data block Size (Bytes)	*Code rate*	*Constituent codes*	*Generator polynomial*
15	~1/2	$(16,11)(16,11)$	$I_x = 0,\ I_y = 0,\ B = 1$
20	~1/2	$(32,26)(32,26)$	$I_x = 13,\ I_y = 13,\ B = 9$
54	~3/5	$(32,26)(32,26)$	$I_x = 2,\ I_y = 8,\ B = 0$
57	~3/5	$(32,26)(32,26)$	$I_x = 2,\ I_y = 7,\ B = 0$
128	~2/3	$(64,57)(64,57)$	$I_x = 25,\ I_y = 25,\ B = 0$
188	~5/7	$(64,57)(64,57)$	$I_x = 25,\ I_y = 10,\ B = 0$
188	~5/7	$(64,57)(64,57)$	$I_x = 17,\ I_y = 17,\ B = 17$
392	~4/5	$(64,57)(64,57)$	$I_x = 1,\ I_y = 1,\ B = 0$

Table 7.2. Recommended TPC Codes

Tables 7.3 and 7.4 present the performance of some recommended codes based on 5 iterations and 4-bit quantization of soft output and QPSK signaling when type 1 interleaver is used.

7.3.2 Advanced Hardware Architectures (AHA)

AHA is the first company that implemented the TPC encoder/iterative decoder for commercial use. The intended applications are *Very Small Aperture Terminal* (VSAT), wireless LAN and broadband wireless link or wireless internet access [154].

For VSAT application where the flexibility in varying packet size used for internet traffic is desirable, TPCs with block structure that can provide such

Code	$(39,32)^2$ $I_x = I_y = 25, B = 0$	$(46,39)^2$ $I_x = I_y = 17, B = 17$	$(63,56)^2$ $I_x = I_y = 1, B = 0$
Rate	0.673	0.711	0.790
Eb/No dB @ 10^{-6} 4/16/64 QAM	3.5/6.5/10.7	3.6/6.6/10.5	3.5/6.6/10.6
Eb/No dB @ 10^{-9} 4/16/64 QAM	4.3/7.5/11.7	4.3/7.8/11.5	4.3/7.5/11.6
Block Size (Information bytes)	1024 (128 bytes)	1504 (188 bytes)	3136 (392 bytes)
Encoder Complexity	10 K gates	10 K gates	10 K gates
Decoder Complexity	Less than 150 K gates	Less than 150 K gates	Less than 150 K gates

Table 7.3. Performance of Recommended Codes

Code	$(16,11)^2$ $I_x = 0,\ I_y = 0,\ B = 1$	$(30,24)(25,19)$ $I_x = 2,\ I_y = 7,\ B = 0$
Rate	0.469	0.608
Eb/No dB @ 10^{-6} 4/16/64 QAM	4.0/6.8/9.8	3.4/6.3/10
Eb/No dB @ 10^{-9} 4/16/64 QAM	5.8/8.8/11.8	4.7/7.5/11.5
Block Size (Information bytes)	120 (15 bytes)	456 (57 bytes)
Encoder Complexity	10 K gates	10 K gates
Decoder Complexity	Less than 150 K gates	Less than 150 K gates

Table 7.4. Performance of Recommended Codes (Cont.)

a flexibility are preferred. Some TPCs with block size of 4000 bits used in satellite link are presented in Table 7.5. The $\frac{E_b}{N_o}$ performance of different TPCs using BPSK modulation over AWGN channel at BER of 10^{-8} is given.

Moreover, research on TPCs for Direct-to-Home Digital Satellite Broadcasting is performed by AHA, [155]. In this application, coding schemes with high bandwidth efficiency are needed. It is stated in [155] that there are two ways to obtain higher bandwidth efficiency. One is to use higher order modulation, the other is to use better FEC. AHA considers different options for coding systems that achieve the goal of a spectral efficiency of 1.85 bits/sec/Hz at $\frac{E_b}{N_o}$ of 7.5 dB. Table 7.6 shows the candidate codes for this specification. The coded $\frac{E_b}{N_o}$s are

Code	*Code Rate*	*Eb/No* dB *@BER* 10^{-8}
$(64,57)(64,57)$	0.793	3.6
$(32,26)(32,26)(4,3)$	0.495	2.4
$(16,11)(16,11)(16,11)$	0.325	1.9

Table 7.5. TPCs used in Satellite Link with Block Size of 4000 Bits

obtained from AHA4540 TPC chip, real hardware implementation at 12 iterations and 4-bits quantization of soft information at BER of 10^{-11}. The modem $\frac{E_b}{N_o}$ that include implementation margins is higher than the coded $\frac{E_b}{N_o}$ due to the imperfect channel estimation. It is observed that 8PSK TCC outperforms QPSK TPC. However, when the implementation issue is considered, the choice of TPC is at least as good as TCC or even more favorable. Therefore, TPCs used with QPSK are further investigated. Table 7.7 lists the choices of TPCs using QPSK modulation with different spectral efficiency ranging from 1.56 to 1.95 bits/sec/Hz. According to AHA [155], the codec AHA4540 can have a throughput as high as 155 Mbps

In wireless LAN applications such as in IEEE 802.11 standard, even though TPC is not specified in the standard, users are welcome to select their choice of FEC. Data sent through the wireless LAN has data packet sizes ranging from 28 bytes to 2.25 kbytes. The code rate flexibility of TPCs is obtained by shortening the codes. Therefore, when the channel is poor, the rate of the code can be adaptively reduced on a block by block basis to achieve better BER. Another advantage of the shortened codes is that one decoder can support different shortened codes resulting from the same mother code, thus only one hardware decoder is needed. Figure 7.8 shows the BER and Packet Error Rate (PER) of the code shortened from a $(64,57) \times (64,57)$ mother code to fit in a 188-byte packet. The (2141,1504) shortened code with code rate of about 0.7 is obtained.

7.3.3 COMTECH EF DATA

In the new satellite transponders, the bandwidth is more scarce than the power. Thus, in designing satellite modems, the system design goal is to find FECs that provide high spectral efficiency. Table 7.8 presents different candidate coding schemes, [156] for $\frac{E_b}{N_o}$ at BER of 10^{-6} and 10^{-8} along with their spectral efficiency. There are three coding schemes employing different modulation techniques, *i.e.* TPC with QPSK, 8PSK and 16QAM, IESS 310 standard coding scheme which uses 8PSK rate 2/3 TCM and RS code and 16QAM RS outer

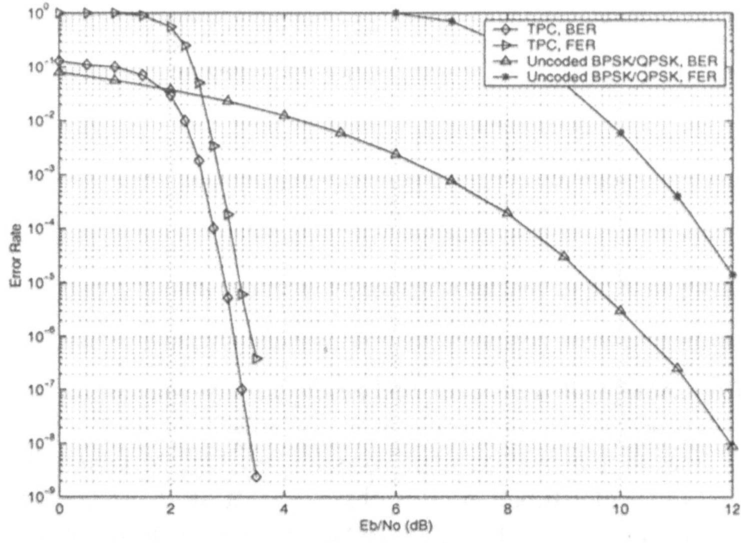

Figure 7.8. Performance of AHA-TPC for Packet Size of 188 bytes

System	Rate	Spectral Efficiency b/s/Hz	Eb/No dB @BER 10^{-11}	Modem Implementation Margin (dB)	Eb/No dB @BER 10^{-11} (Implementation)
8PSK DVB-DSNG Viterbi/Reed-Solomon	0.61	1.84	8.6	1	9.6
8PSK TCC with RS outer (48 Kbits block)	0.61	1.83	7.1	1.3	8.4
8PSK TPC $(64, 57)(32, 26)(8, 7)$	0.63	1.90	7.3	1.3	8.6
QPSK TPC $(128, 120)(129, 127)$	0.92	1.845	7.8	0.6	8.4

Table 7.6. Possible Coding Schemes for 1.85 bps/Hz Spectral Efficiency

code and convolutional inner code. It is noted that by using TPCs, latency can be reduced by 70% as compared to that of the concatenation of the RS and convolutional codes. Furthermore, the problem of threshold effect, that is, extreme sensitivity to changes in $\frac{E_b}{N_o}$ is less noticable in the case of TPCs. The choice of TPCs can be made based on the situations, *i.e.* power limitation or bandwidth limitation. Table 7.9 shows the $\frac{E_b}{N_o}$ of the TPCs with different code rates and different modulation schemes such as QPSK/OQPSK, BPSK and 8PSK at BER of 10^{-6} and 10^{-9} together with their speed.

System	Rate	Spectral Efficiency b/s/Hz	Eb/No dB @BER 10^{-11}	Modem Implementation Margin (dB)	Eb/No dB @BER 10^{-11} (Implementation)
QPSK TPC $(128,127)(129,127)$	0.98	1.95	9.8	0.6	10.4
QPSK TPC $(128,120)(129,127)$	0.92	1.84	7.9	0.6	8.4
QPSK TPC $(128,120)(129,120)$	0.87	1.74	6.5	0.6	7.1
QPSK TPC $(64,57)(65,57)$	0.79	1.56	5.8	0.6	6.4

Table 7.7. Performance of TPCs using QPSK Modulation Scheme

Mode	*Eb/No dB @BER 10^{-6} Guaranteed (Typical)*	*Eb/No dB @BER 10^{-8} Guaranteed (Typical)*	*Spectral Efficiency*	*Occupied Bandwidth for 1Mbps Carrier*
QPSK Rate 3/4 Turbo	3.8 (3.3)	4.4 (3.9)	1.50 b/Hz	793 kHz
8PSK Rate 2/3 TCM and Reed-Solomon (IESS-310)	6.5 (6.2)	6.9 (6.6)	1.82 b/Hz	654 kHz
8PSK Rate 3/4 Turbo	7.7 (7.3)	8.3 (7.8)	2.25 b/Hz	529 kHz
16QAM Rate 3/4 Turbo	7.8 (7.4)	8.6 (8.2)	3.00 b/Hz	396 kHz
16QAM Rate 3/4 Viterbi/Reed-Solomon	8.1 (7.5)	8.6 (8.0)	2.73 b/Hz	435 kHz
16QAM Rate 7/8 Viterbi/Reed-Solomon	9.5 (9.0)	10.1 (9.5)	3.18 b/Hz	374 kHz

Table 7.8. Possible Coding Scheme for Use in Satellite Modem

7.3.4 Turbo Concept

TPCs proposed by Turbo Concept are provided as *Intellectual Property* (IP) cores [157]. Hamming codes, BCH codes with double errors correcting capability and parity check code are used as the component codes of product codes. Bit rates ranging from 7 to 25 Mbps are achieved at 5 decoding iterations. Table

Modulation type	*QPSK/OQPSK*	*QPSK/OQPSK*	*BPSK*	*8PSK*
Rate	1/2	3/4	21/44	3/4
Eb/No dB @BER 10^{-6}	3.0	3.9	2.9	7.0
Eb/No dB @BER 10^{-9}	3.8	4.7	3.7	8.0
Data Rate Range (Kbps)	4.8-2386	7.2-3750	2.4-1193	384-5000

Table 7.9. Performance and Data Rate of TPC using Different Modulation Scheme

7.10 shows that performance of different TPCs at BER of 10^{-5} and 10^{-8} using QPSK modulation.

Product Code	*Rate*	*Eb/No dB* @BER 10^{-5}	*Eb/No dB* @BER 10^{-8}
$(32, 26)(32, 26)$	0.660	2.9	3.6
$(32, 21)(32, 21)$	0.431	2.4	N.A.
$(64, 57,)(64, 57)$	0.793	3.2	3.6
$(64, 51)(64, 51)$	0.635	2.6	2.9
$(128, 120)(128, 120)$	0.879	3.8	4.2
$(128, 113)(128, 113)$	0.779	3.3	3.4
$(256, 247)(256, 247)$	0.931	4.5	4.8
$(256, 239)(256, 239)$	0.872	4.0	N.A.

Table 7.10. Performance of BCH-TPCs with Different Block Size

7.3.5 Paradise Data Com

TPCs are used in Paradise Data Com Satellite modems [158]. Use of turbo code, instead of the conventional concatenated coding scheme, gives the system providers, the opportunity to reduce both the power and bandwidth requirement for a given bit rate and bit error rate. This saving can be translated into lower Earth station cost (smaller SSPAs, smaller antennas), lower space segment cost, or higher throughput depending on the customer demands

Table 7.11 shows the performance of different rates of TPC codes used in the Paradise Data Com's modems.

Rate	Eb/No *dB* @*BER* 10^{-5}	Eb/No *dB* @*BER* 10^{-8}
1/2	2.9	3.7
3/4	3.7	4.5
7/8	6.6	8.3

Table 7.11. Performance of TPCs used in Paradise Data Com's Satellite Modem

7.4. Summary

In this chapter, we presented information related to the performance of BTCs and their applications. The results presented in this chapter are from academic publications, standard proposal as well as companies' white papers. It is shown that BTC's performance is comparable with TCCs and outperforms them in some cases. This shows that overall, the iterative decoding technique is the key to the excellent performance of turbo codes, rather than the constituent codes. One could choose either TCCs or BTCs depending on the applications. BTCs perform well in the moderate to high SNRs because the effect of error floor is less. As BTCs have more advantage when a high rate code is used, they are suitable for commercial applications in wireless and satellite communications. Moreover, their code rate flexibility and block structure which can be adaptively changed on a block by block basis makes BTCs attractive for packet transmission. Furthermore, BTCs can be decoded using the algebraic-based algorithm, which has low complexity and is easy to implement. This allows the system to operate at speeds as high as 155 Mbps.

Chapter 8

IMPLEMENTATION ISSUES

As discussed earlier, tubo codes have an amazing error correcting capability and are, therefore, very attractive for many applications. In this chapter, we address some of the implementation issues. The complexity of a turbo-decoder is much higher than the complexity of the encoder. Thus, we put emphasis on the decoder. Low cost and low power consumption are extremely important issues for turbo decoder implementation. Consequently, fixed-point arithmetic and fixed-point implementation are unavoidable issues. The Max-Log-MAP algorithm discussed in this book is simple enough and performs very close to the MAP algorithm. So, it is a good trade-off between the complexity and performance. In this chapter, the effect of input data quantization for TCC and BTC are presented. Moreover, the effect of correction term in Max-log-MAP algorithm is discussed.

So far, we have assumed perfect carrier phase and channel SNR estimation in our simulations. However, this does not truly represent a practical system where *channel impairments* caused by the noise and attenuation occur. Some examples of channel impairments are carrier phase offset and channel SNR mismatch, i.e., a difference between the assumed and actual values of the phase and the SNR. In this chapter, the effect of channel impairements on turbo codes are presented. We will also discuss hardware implementation for turbo codes on FPGA, ASIC and DSP.

8.1. Fixed-point Implementation of Turbo Decoder

Algorithm used in a turbo decoder are usually specified in the floating point domain. Fixed point number representation is mandatory for most target architectures, thus quantization is a necessary step towards an actual implementation [61].

Quantization is the process of representing the data with one or a few bits of precision. In channel coding, the channel symbols are corrupted by the channel noise and interference in the digital communication system. Applying the quantizer at the receiver makes the channel-decoder work with finite precision or with fixed-point arithmetic.

There are a few strategies for turbo-decoder quantization depending on the decoding algorithm. Input data quantization is an important issue. The effect of the finite accuracy of the internal values is addressed in [94] [95]. For MAP, Log-MAP, Max-Log-MAP and SOVA decoding algorithms, one strategy is to use uniform quantization of all signals [96][97] and a systematic approach towards an internal quantization scheme for a 4-state turbo-decoder with finite accuracy of the input data is discussed in [98]. The first investigation of combined bit-width optimization of the input data and the internal data for an 8-state turbo-decoder based on UMTS parameters is discussed in [99].

The objective of data quantization may differ depending on the implementation platform. The primary goal for a software implementation is to find a fixed-point model that corresponds to the given bit-width of the DSP. Further bit-width minimization can reduce the switching activity and has thus influence on the power consumption. On the other hand, the primary goal for a dedicated hardware implementation is to choose all bit-widths as small as possible, resulting in a reduction of area and power consumption. Hence, an optimized quantization has a major impact on the implementation cost. Both 3-bit and 4-bit quantization are discussed in this chapter.

The strategy of turbo-decoder quantization described in [99], is optimal for MAP or Log-MAP decoding algorithm with m-bit input samples, and is impossible in the case of a Max-Log-MAP decoding algorithm implementation because of the approximation.

$$E_{i=1}^{k} x_i = \ln \sum_{i=1}^{k} e^{x_i} \approx \max(x_i). \tag{8.1}$$

In Section 8.1, different schemes of input data quantization for an 8-state double-binary CRSC code is designed for a wide range of coding rates. The system model for turbo-decoder quantization is depicted in Figure 8.1. The QPSK demodulator, quantizer and turbo decoder are combined together in the simulation. Also only the channel output, the input data of the decoder, is quantized. In Section 8.1.2, we will present the effect of the input data quantization on two and three dimensional-RM turbo codes.

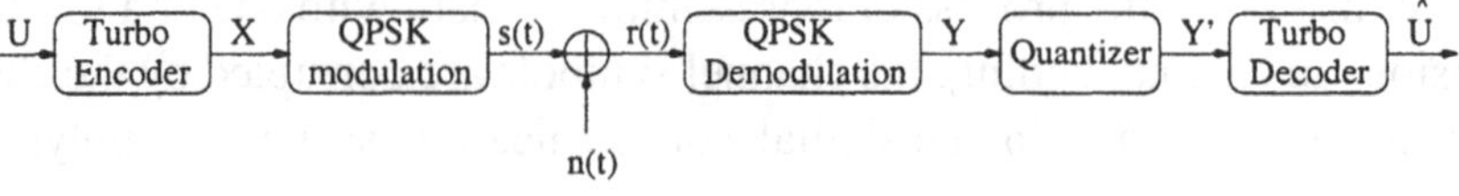

Figure 8.1. System Model for Quantization

8.1.1 Input Data Quantization for DVB-RCS Turbo Codes

In practical systems, the received channel symbols should be quantized with one or a few bits of precision in order to reduce the complexity of the turbo decoder. The usual quantization precision is three bits as introduced in [100]. For the QPSK modulation and an AWGN channel, the received values are corrupted with a Gaussian distribution (see Figure 8.2) around the transmitted symbols {-1, 1}.

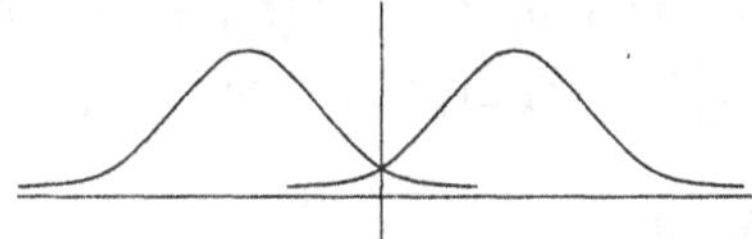

Figure 8.2. The Distribution of the Transmitted Symbols

More than 99% of the observed values are in the range of [-4, 4], *i.e.*, $\pm 1 \pm 3\sigma$[1] This dynamic-range is reasonable and can be represented by 3 bits in a uniform quantization. Using more bits results in higher complexity, but less degradation in the performance. Let's use a uniform, 3-bit quantizer having the input/output relationship shown in the Figure 8.3, where D is the size of the quantization step.

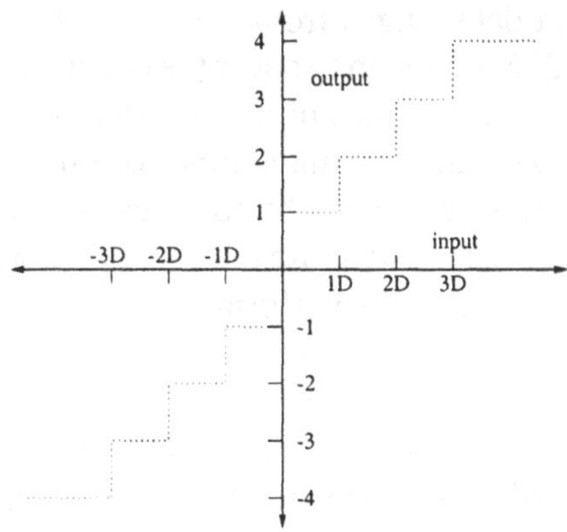

Figure 8.3. Quantizer Model in 3-bit

The selection of the quantization step size is an important consideration because it can have a significant effect on the performance. The step size, D, can be chosen as a fixed value for different coding rates, for example, $D = 11$,

[1] For an $N(\mu, \sigma)$ Gaussian random variable, the distribution function is given by $F(x) = Q\left(\frac{x-\mu}{\sigma}\right) = 1 - Q\left(\frac{\mu-x}{\sigma}\right)$. Therefore $P[\mu - 3\sigma < x < \mu + 3\sigma] = Q(-3) - Q(3) = 1 - 2Q(3) = 0.9974$.

or calculated according to the formula $D = f \cdot \sigma = f \cdot \sqrt{1/(2 \cdot (E_s/N_0))}$, where E_s/N_0 is the energy per symbol to noise density ratio, and f is a factor depending on the code rate.

Denote the channel symbols at the receiver as $y_k^{s,I}$, $y_k^{s,Q}$, $y_k^{p,I}$, $y_k^{p,Q}$ as in Chapter 3. Since the branch transition probability

$$\begin{aligned} & \ln \gamma_k^i(S_{k-1}, S_k) \\ = & \tfrac{1}{2} L_c \cdot [y_k^{s,I} \cdot x_k^{s,I}(i) + y_k^{s,Q} \cdot x_k^{s,Q}(i)] + \ln P(d_k) + K \\ = & +\tfrac{1}{2} L_c \cdot [y_k^{p,I} \cdot x_k^{p,I}(i, S_{k-1}, S_k) + y_k^{p,I} \cdot x_k^{p,Q}(i, S_{k-1}, S_k)] \end{aligned} \tag{8.2}$$

and the extrinsic information

$$\begin{aligned} L_i^e(\hat{d}_k) = & L_i(\hat{d}_k) - \frac{1}{2} L_c \cdot [y_k^{s,I} \cdot x_k^{s,I}(i) + y_k^{s,Q} \cdot x_k^{s,Q}(i)] \\ & + \frac{1}{2} L_c \cdot [y_k^{s,I} \cdot x_k^{s,I}(0) + y_k^{s,Q} \cdot x_k^{s,Q}(0)] - \ln \frac{P[d_k = i]}{P[d_k = 0]} \end{aligned} \tag{8.3}$$

relate to the received channel symbols, where L_c is the reliability value of the channel and $i = 1, 2, 3$, they are quantized after being multiplied by $\frac{1}{2} \cdot L_c$.

The input data of the decoder is multiplied by 2^5 and truncated at -128 and 127 according to the limited dynamic-range of the received channel values to [-4, 4]. Since the input data ranges from -128 to 127, an 8-bit look-up table with indices from 0 to 256 is enough to cover the occurring values of the decoder input data. Thus, all the calculations of the decoder are integer. This approach is very flexible since, we can vary the number of quantization bits and make them as small as possible according to the complexity level that can be afforded. We can also vary the step sizes and chose the value giving the best performance.

Simulation results show that the performance of a decoder with 3-bit quantization is very sensitive to the step size chosen. In Figure 8.4 and Figure 8.5, the selection of step size depends on the code rate and the performance of the adaptive step size is better than that of the fixed step size. The parameters of fixed step size and adaptive step size are presented in Table 8.1.

To achieve better performance, 4-bit quantization is used in the same way as shown in Figure 8.6. There are two kinds of step sizes, fixed and adaptive step sizes.

For higher code rates, we have to modify the step size of the quantizer. For adaptive step size, D can be calculated according to the formula,

$$\begin{aligned} D &= f \cdot \sigma_s \\ &= f \cdot \sqrt{1 + \sigma^2} \\ &= f \cdot \sqrt{1 + 1/(2 \cdot (E_s/N_0))} \end{aligned} \tag{8.4}$$

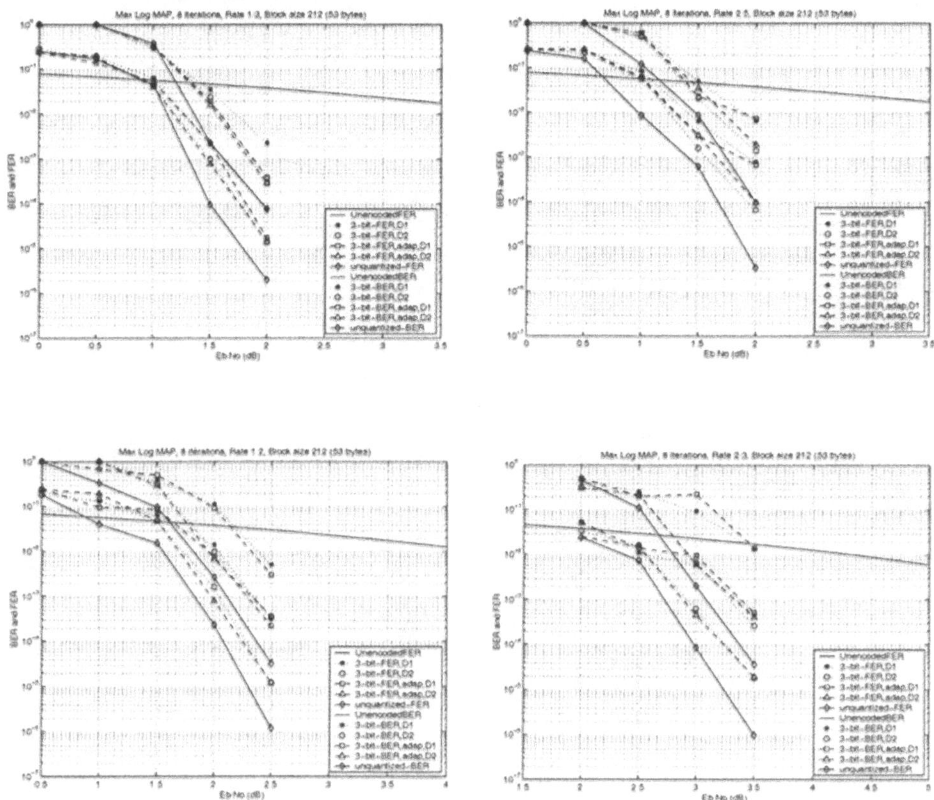

Figure 8.4. 3-bit Quantization. Code Rate=1/3, 2/5, 1/2, 2/3. The Parameters of Decision Level refer to Table 8.1.

where E_s/N_0 is the energy per symbol to noise density ratio and f is a factor depending on the coding rate.

Figure 8.7 and Figure 8.8 show that 4-bit quantization of the channel output data is a reasonable compromise between implementation complexity and degradation of the decoding performance. The performance of the adaptive step size is better than that of a fixed step size and very close to the unquantized performance. So, we can say that there is no degradation in the decoding performance.

8.1.2 Input Data Quantization for BTC

The study of input data quantization for two and three dimensional-RM turbo codes is done in [160]. The approach is similar to the one presented in Section 8.1.1. That is, uniform quantization is used where the limiting range of received sequence is $(-1-3\sigma, +1+3\sigma)$ where -1 and +1 are the possible transmitted signal levels and σ is the variance of the AWGN noise. This range covers more

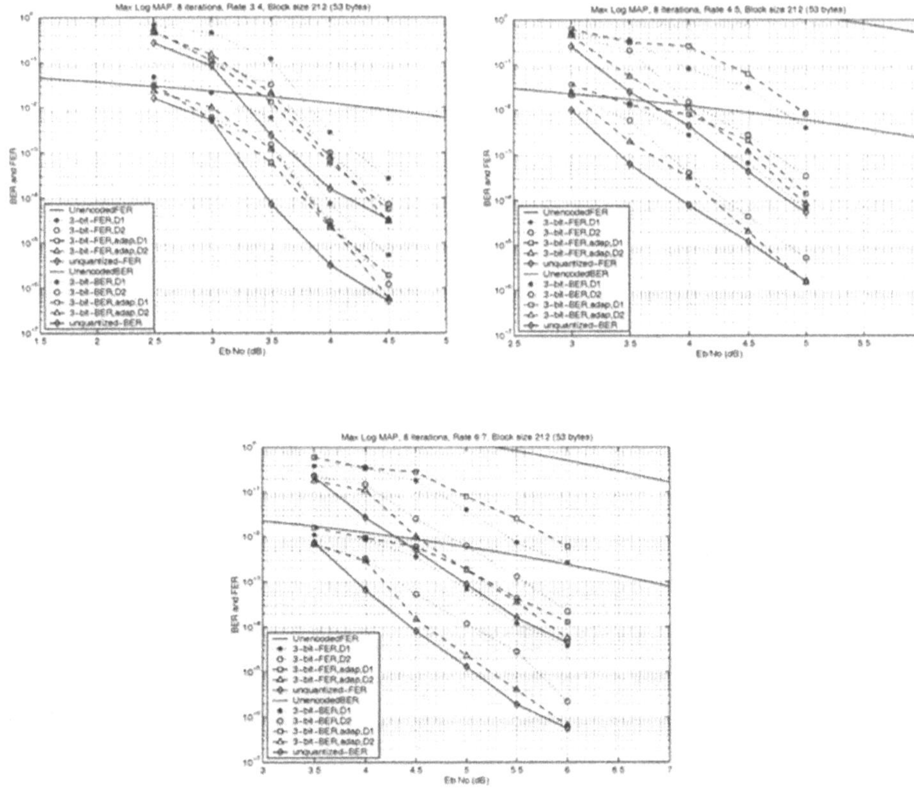

Figure 8.5. 3-bit Quantization, Code Rate: 3/4, 4/5, 6/7. The Parameters of Step Size refer to Table 8.1.

than 99% of all possible received data points. In [160], σ is derived based on $\frac{E_b}{N_o} = 5$dB. The investigation of the effect of quantization bits is illustrated in Figure 8.9.

MAP algorithm and Max-log-MAP algorithm with correction factor, denoted as Max-corr, are considered. In the case of 4-bit quantization with Max-corr, performance degradation is about 0.1 and 0.2 dB at BER of 10^{-5} when compared to 5-bit quantization with Max-corr and real valued MAP algorithm, respectively. The higher the number of quantization bits, the more complex is the decoder. Therefore, 4-bit quantization seems to provide a good compromise between the performance loss and the decoding complexity.

Figure 8.10 depicts the performance comparison of three-dimensional RM turbo codes using Max-corr algorithm with 4-bit quantization with real-valued MAP and Max-corr algorithm. Performance loss of 0.3 dB at BER of 10^{-5} is observed as compared to the floating point MAP algorithm.

Code rate	Fixed step size	Adaptive step size $\sigma = \sqrt{1/2 * (E_s/N_0))}$
$R = 1/3$	$D1 = -128, -31, -21, -11, 0, 11, 21, 31, 127$	$D1 = 0.51 * 32 * \sigma$
	$D2 = -128, -56, -36, -18, 0, 18, 36, 56, 127$	$D2 = 0.53 * 32 * \sigma$
$R = 2/5$	$D1 = -128, -31, -21, -11, 0, 11, 21, 31, 127$	$D1 = 0.45 * 32 * \sigma$
	$D2 = -128, -56, -36, -18, 0, 18, 36, 56, 127$	$D2 = 0.55 * 32 * \sigma$
$R = 1/2$	$D1 = -128, -31, -21, -11, 0, 11, 21, 31, 127$	$D1 = 0.5 * 32 * \sigma$
	$D2 = -128, -63, -42, -21, 0, 21, 42, 63, 127$	$D2 = 0.55 * 32 * \sigma$
$R = 2/3$	$D1 = -128, -31, -21, -11, 0, 11, 21, 31, 127$	$D1 = 0.5 * 32 * \sigma$
	$D2 = -128, -63, -42, -21, 0, 21, 42, 63, 127$	$D2 = 0.55 * 32 * \sigma$
$R = 3/4$	$D1 = -128, -31, -21, -11, 0, 11, 21, 31, 127$	$D1 = 0.725 * 32 * \sigma$
	$D2 = -128, -65, -45, -21, 0, 21, 45, 65, 127$	$D2 = 1.0 * 32 * \sigma$
$R = 4/5$	$D1 = -128, -31, -21, -11, 0, 11, 21, 31, 127$	$D1 = 0.5 * 32 * \sigma$
	$D2 = -128, -63, -42, -21, 0, 21, 42, 63, 127$	$D2 = 1.0 * 32 * \sigma$
$R = 6/7$	$D1 = -128, -31, -21, -11, 0, 11, 21, 31, 127$	$D1 = 0.5 * 32 * \sigma$
	$D2 = -128, -63, -41, -20, 0, 20, 41, 63, 127$	$D2 = 1.0 * 32 * \sigma$

Table 8.1. Parameters of Fixed Step Size and Adaptive Step Size (3-bit Quantization)

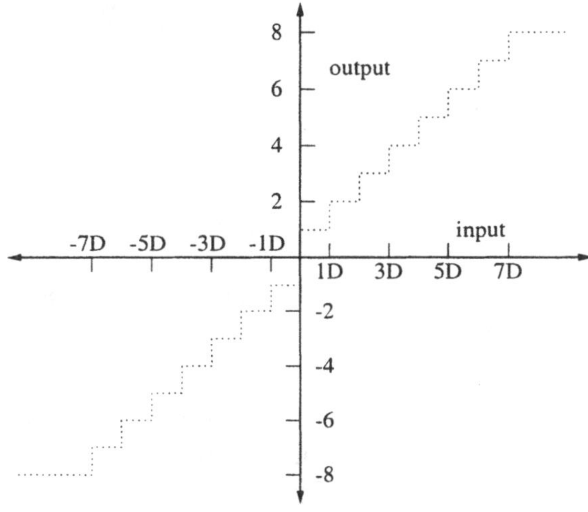

Figure 8.6. 4-bit Quantization Level

8.2. The Effect of Correction Term in Max-Log-MAP Algorithm

Max-Log-MAP algorithm is derived from the Log-MAP algorithm by approximating $\ln \sum_{i=1}^{k} e^{x_i}$ with $\max(x_i)$. This approximation results in some degradation in the performance compared to that of MAP algorithm. Here we discuss the correction term required to be added so that the performance of the Max-Log-MAP algorithm approaches that of the MAP algorithm.

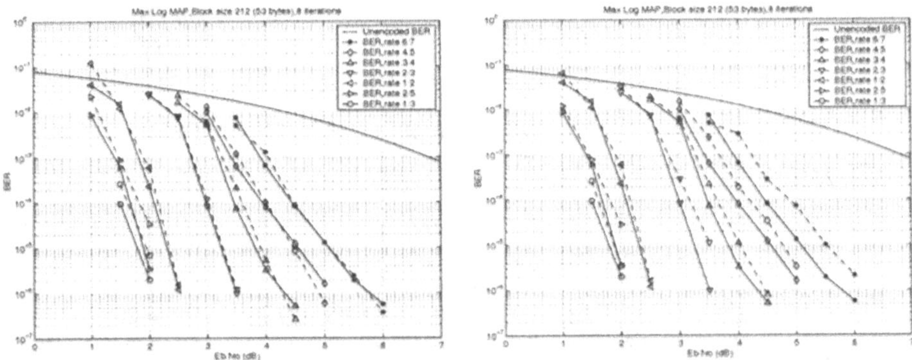

Figure 8.7. 4-bit Quantization with Adaptive Decision Level. The Solid lines are unquantized and the dashed lines are quantized with 4-bit.

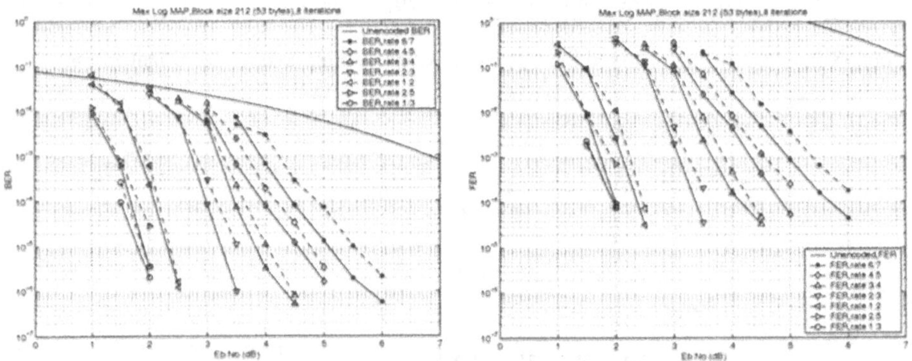

Figure 8.8. 4-bit Quantization with Fixed Decision Level. The Solid lines are unquantized and the dashed lines are quantized with 4-bit.

For a binary convolutional turbo code, the correct function in Equation (2.50) is $\ln(1+e^{-|x-y|})$. Table 8.2, is the 3-bit look-up table containing the values of correction term for different values of $|x-y|$.

$\|x-y\|$	0.00	0.25	0.50	0.75	1.00	1.25	1.50	1.75	>2.00
$\ln(1+e^{-\|x-y\|})$	0.69	0.58	0.47	0.39	0.31	0.25	0.20	0.16	0

Table 8.2. Look-up Table for Correction Term in Binary Convolutional Turbo Code

A 1-bit approximation is given by

$$\ln(1+e^{-|x-y|}) = \begin{cases} 3.0/8.0 & if\ x < 2.0 \\ 0 & otherwise \end{cases} \tag{8.5}$$

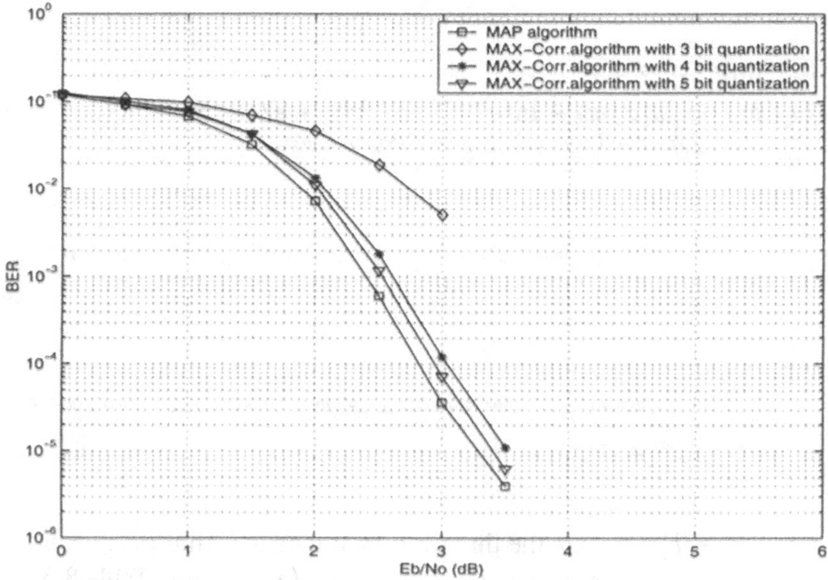

Figure 8.9. The Effect of Number of Quantization Bits on RM(32, 26)^2

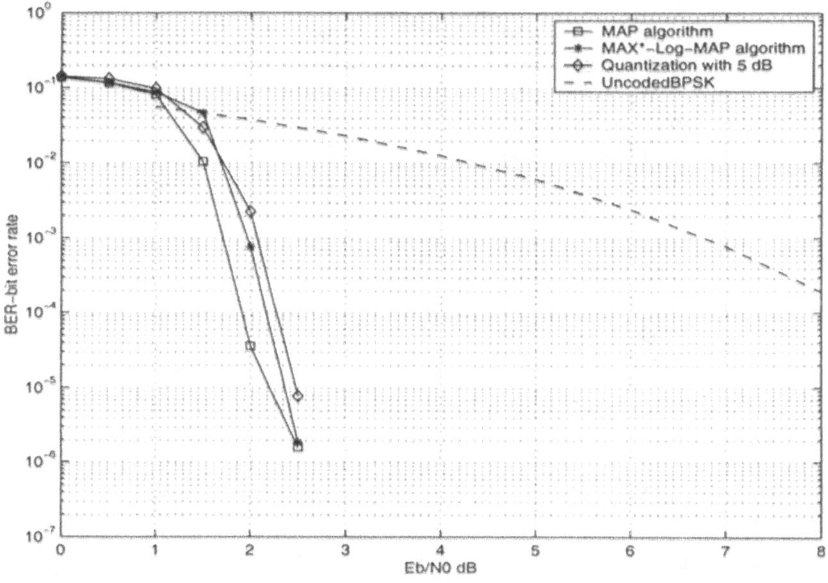

Figure 8.10. The Effect of Channel Input Quantization on RM(32, 26)^3

The performance degradation due to the application of a Max-Log-MAP algorithm is less significant in the case of double-binary codes (less than 0.1dB) than in the case of binary codes (0.3 to 0.4 dB) [36]. The four term sum is written as,

$$\begin{aligned} & \ln(e^x + e^y + e^z + e^w) \\ = \; & \max(x, y, z, w) + \ln(e^{x-\max(x,y,z,w)} + e^{y-\max(x,y,z,w)} \\ & + e^{z-\max(x,y,z,w)} + e^{w-\max(x,y,z,w)}) \end{aligned} \tag{8.6}$$

where

$$0 < \ln(e^{x-\max(x,y,z,w)} + e^{y-\max(x,y,z,w)} + e^{z-\max(x,y,z,w)} + e^{w-\max(x,y,z,w)}) \leq \ln 4 \tag{8.7}$$

Define

$$\begin{aligned} & \ln(e^{x-\max(x,y,z,w)} + e^{y-\max(x,y,z,w)} + e^{z-\max(x,y,z,w)} + e^{w-\max(x,y,z,w)}) \\ = \; & \ln(1 + e^{-|a|} + e^{-|b|} + e^{-|c|}) \end{aligned} \tag{8.8}$$

where $-|a|, -|b|, -|c|$ are the three values among $x - \max(x, y, z, w)$, $y - \max(x, y, z, w)$, $z - \max(x, y, z, w)$, or $w - \max(x, y, z, w)$. Table 8.3, gives the values $\ln(1 + e^{-|a|} + e^{-|b|} + e^{-|c|})$ for different values of $|a| = |b| = |c|$.

$\|a\|$	$\|b\|$	$\|c\|$	$\ln(1 + e^{-\|a\|} + e^{-\|b\|} + e^{-\|c\|})$
0.00	0.00	0.00	1.386
0.25	0.25	0.25	1.204
0.50	0.50	0.50	1.037
0.75	0.75	0.75	0.883
1.00	1.00	1.00	0.744
1.25	1.25	1.25	0.620
1.50	1.50	1.50	0.512
1.75	1.75	1.75	0.420
2.00	2.00	2.00	0.341
2.50	2.50	2.50	0.220
3.00	3.00	3.00	0.139

Table 8.3. Look-up Table for Correction Term

To avoid increased complexity, a two level quantizer given by

$$\ln(1+e^{-|a|}+e^{-|b|}+e^{-|c|}) = \begin{cases} 5.0/8.0 & if \;\; \max(|a|, |b|, |c|) < 2.0 \\ 0 & otherwise \end{cases} \tag{8.9}$$

can be used.

Figure 8.11 shows the performance of double-binary CRSC codes when a 1-bit look-up table is used. Simulation results show that using a 1-bit look-up table does not affect the performance of the code much.

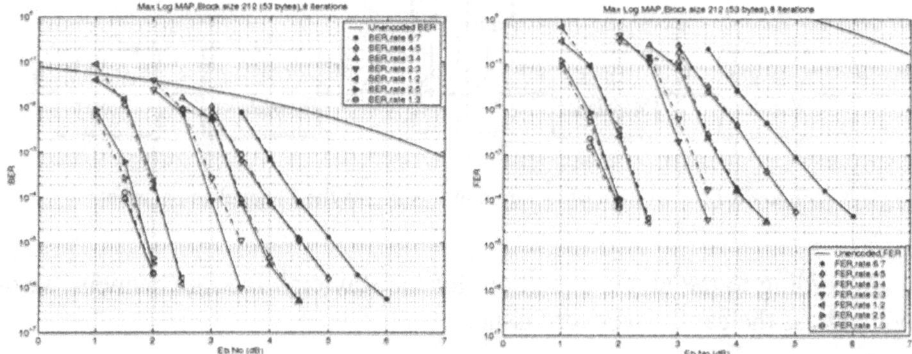

Figure 8.11. With Correct Coefficient: Two Level Look-up Table. The dashed lines are the performances with correction coefficient.

8.3. Effect of Channel Impairment on Turbo Codes

In this section, we present the effect of channel SNR mismatch and input quantization on the RM-turbo codes. Additionally, the effect of phase offset on the shortened RM-turbo codes is discussed. Then, the impact of preamble size on the performance of the RM-turbo codes is also shown.

8.3.1 System Model for the Investigation of Channel Impairments

The channel model used for investigating the channel impairments is illustrated in Figure 8.12. The received signal can be written as,

$$r(t) = \alpha(t)e^{j\Delta\phi}s_i(t) + n(t) \tag{8.10}$$

where $\alpha(t)$ is a Rayleigh process that satisfies $E(\alpha_i^2) = 1$, $\Delta\phi$ is a random phase offset uniformly distributed over $[-a, a]$, $(\Delta\phi \sim U\ [-a, a])$ where a depends on the variance of $\Delta\phi$, and $n(t)$ is a Gaussian noise process with two-sided power spectral density $N_o/2$. We assume that QPSK modulation is used and $s_i(t)\ : i = 1, 2, 3, 4$ is the modulated waveform for the symbol s_i .

For the case of an AWGN channel we consider two cases :

AWGN channel : without phase offset ; $\alpha(t) = 1$ and $\Delta\phi = 0$
with phase offset ; $\alpha(t) = 1$ and $\Delta\phi \sim U\ [-a, a]$

For the Rayleigh fading channel, we assume that slow fading is applied in such a way that at the receiver, the phase can be recovered using standard techniques and the coherent detection can be used.

Fading channel: $\alpha(t)$ is Rayleigh process ; $\Delta\phi = 0$

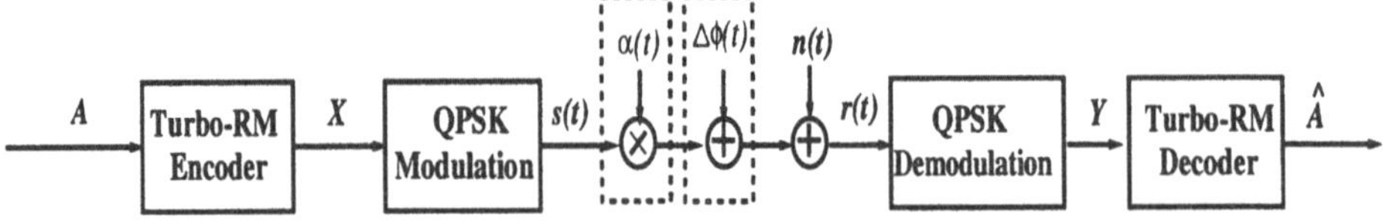

Figure 8.12. System Model used to Investigate the Channel Impairments

According to the above condition, for the Rayleigh fading channels, we do not address the carrier recovery issue.

8.3.2 Channel SNR Mismatch

The knowledge of the channel SNR plays an important role in the iterative MAP decoding through the soft-output calculation. Thus, the incorrect estimation of the channel SNR will affect the performance of the turbo codes.

The log-likelihood ratio of bit x conditioned on the corresponding received signal y, $L(x \mid y)$ at the detector is given by

$$\begin{aligned} L(x \mid y) &= \ln \frac{P(x=+1 \mid y)}{P(x=-1 \mid y)} \\ &= \ln \left(\frac{p(y \mid x=+1)}{p(y \mid x=-1)} \cdot \frac{P(x=+1)}{P(x=-1)} \right) \\ &= \ln \frac{e^{\left(-\frac{E_s}{N_o}(y-a)^2\right)}}{e^{\left(-\frac{E_s}{N_o}(y+a)^2\right)}} + \ln \frac{P(x=+1)}{P(x=-1)} \\ &= L_c \cdot y + L(x) \end{aligned} \tag{8.11}$$

where $L_c = 4 \cdot a \cdot \frac{E_s}{N_o}$ is called the reliability value of the channel, a is the fading attenuation. For a Gaussian channel a equals 1 and $L(x)$ is the *a priori* value. The channel SNR mismatch affects the value of L_c used in the iterative decoding, for instance, if the estimated $\frac{E_b}{N_o}$, i.e., $\widetilde{\frac{E_b}{N_o}} = \frac{1}{R_c} \cdot \widetilde{\frac{E_s}{N_o}}$, where R_c is the code rate, is underestimated by d dB. Then, L_c at the receiver is given by

$$\begin{aligned} L_c &= 4 \cdot a \cdot R_c \cdot \widetilde{\frac{E_b}{N_o}} \\ &= 4 \cdot a \cdot R_c \cdot \left(\frac{E_b}{N_o} / 10^{\frac{d}{10}} \right) \end{aligned} \tag{8.12}$$

We investigate the effect of the channel SNR mismatch in terms of $\frac{E_b}{N_o}$ since the performance evaluation is usually done in terms of $\frac{E_b}{N_o}$. However, in the case of channel SNR estimation, the $\frac{E_s}{N_o}$ is directly considered.

It is obvious that in the above formulae, the turbo decoder that uses the iterative MAP decoding requires the knowledge of the channel SNR or $\frac{E_s}{N_o}$. Where the $\frac{E_s}{N_o}$ is calculated as,

$$\frac{E_s}{N_o} = \frac{channel\ signal\ variance}{noise\ variance} \tag{8.13}$$

Some estimation methods of the channel SNR are studied and presented in [150], [151], where the first one uses the polynomial approximation of the channel SNR obtained from the mean and the variance of the received bits and the second one obtains the channel SNR from the variation of the extrinsic information at each iteration. In [152], the hard decision from turbo decoder and received sequences are used to estimate the noise variance. In this book, we use the channel SNR estimation algorithm of [152] to calculate the noise variance, due to its simplicity and good performance.

8.3.2.1 Simulation Results. Figure 8.13 shows the BER versus channel SNR mismatch for RM $(8,4)^2$ - turbo code at different $\frac{E_b}{N_o}$. It is shown that the performance in terms of BER is very little degraded for a channel SNR mismatch of -6 dB and less, otherwise there is no degradation observed.

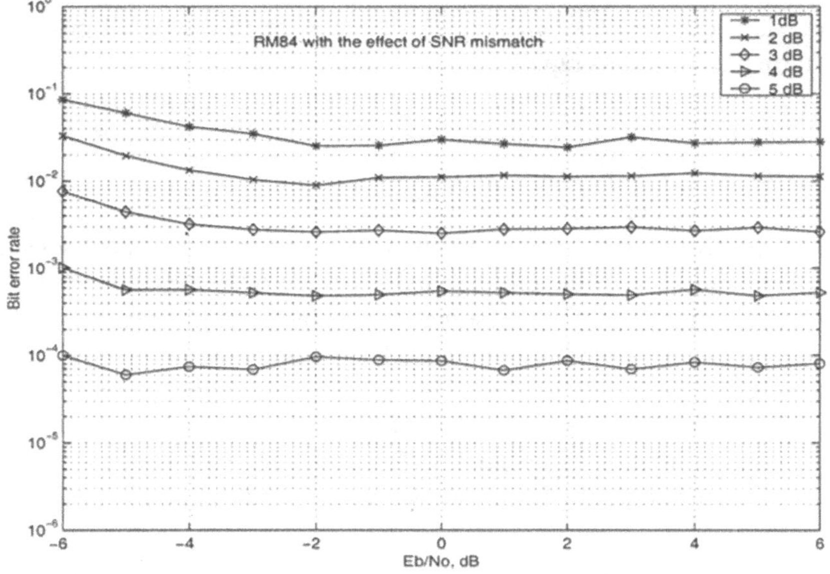

Figure 8.13. Effect of Channel SNR Mismatch on Performance of a RM $(8,4)^2$-turbo Code

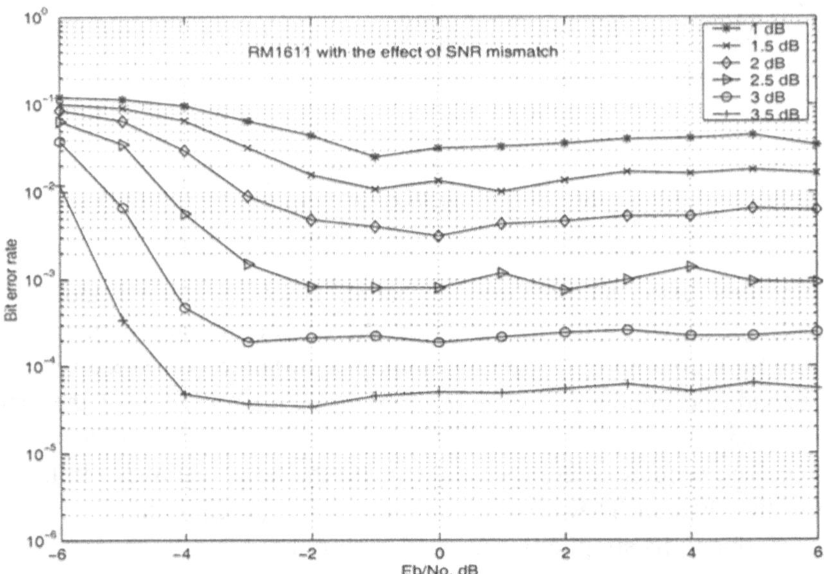

Figure 8.14. Effect of Channel SNR Mismatch on Performance of a RM $(16, 11)^2$-turbo Code

The BER versus channel SNR mismatch for RM $(16, 11)^2$-turbo code at different $\frac{E_b}{N_o}$ is presented in Figure 8.14, where the degradation on the performance is observed when the channel SNR mismatch is less than -3 dB. Similarly, Figure 8.15 illustrates the BER versus the channel SNR mismatch for RM $(32, 26)^2$ - turbo code at different $\frac{E_b}{N_o}$, where at -2 dB or less of channel SNR mismatch, the performance degrades rapidly.

From these three figures, it can be seen that the higher the $\frac{E_b}{N_o}$, the more the tolerance to channel SNR mismatch, for example in Figure 8.14, the start points of the performance degradation are -4, -3 and -2 dB or less at $\frac{E_b}{N_o}$ of 3.5, 3 and 2.5 dB, respectively. It is shown that the RM-turbo codes are more sensitive to the underestimation of the channel SNR than the overestimation of it. The reason is that, in the case of underestimation, the factor used for calculating the soft-output is smaller than it should be so less information could be extracted and transfered between the two decoders resulting in no improvement from the iterative decoding. Also the longer the code length, the more significant is the effect of underestimation of channel SNR mismatch. This is due to the fact that the longer the code, the larger interleaver size and the more powerful the decoding process.

In [150], the effect of SNR mismatch is investigated on a rate $\frac{1}{3}$ parallel TCC with helical interleaver. Block size of 420 bits and log-MAP decoding algorithm are considered. The work in this paper is presented before our investigation on

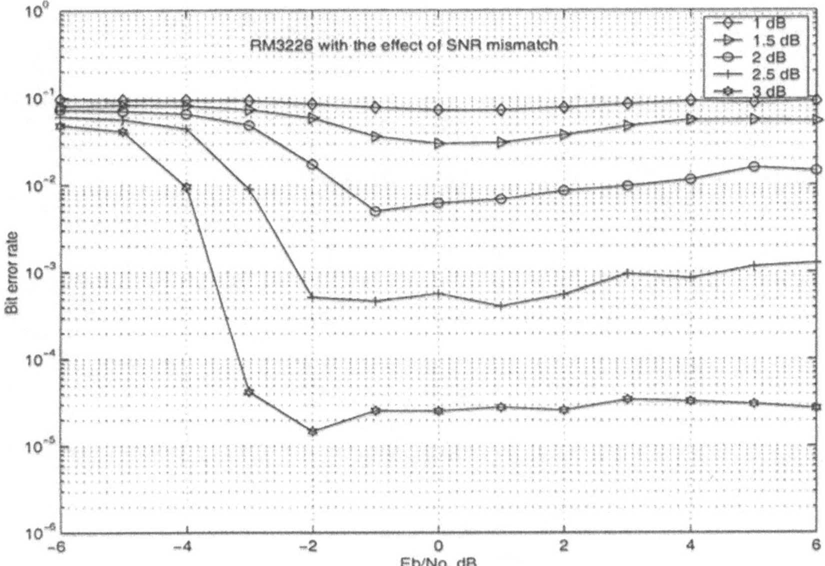

Figure 8.15. Effect of Channel SNR Mismatch on Performance of a RM $(32, 26)^2$-Turbo Code

RM-turbo codes. We found similar behavior as their result and it is depicted in Figure 8.16. The code is more tolerable of the overestimation of SNR than the underestimation of SNR. Moreover, the effect of SNR mismatch is studied on a serial TCC with the rate $\frac{1}{3}$ in [152]. A 2048-bit long S-random interleaver is used. Figure 8.17 shows that the serial TCC performs poorly when the SNR is off from the true SNR regardless of whether it is higher or lower. The reason for such a behavior might be that SNR is estimated only at the inner decoder, whereas SNR is estimated from both decoders in the case of parallel turbo code. It should be noted that RM-turbo code is a parallel concatenated code.

The BER versus $\frac{E_b}{N_o}$ of RM $(32, 26)^2$ - turbo code with and without variance estimations on Gaussian channel is shown in Figure 8.18. It is shown that the estimation algorithm performs well. The BER versus $\frac{E_b}{N_o}$ of RM-turbo code with and without variance estimations on Rayleigh-fading channel is presented in Figure 8.19. Performance is better when the variance estimation is performed rather than when constant channel reliability is assumed, even though perfect channel SNR is used. This is due to the time variance of the fading channel. However, the improvement is modest in the case of RM $(8, 4)^2$ and $(16, 11)^2$ turbo codes because they are less sensitive to the channel SNR mismatch.

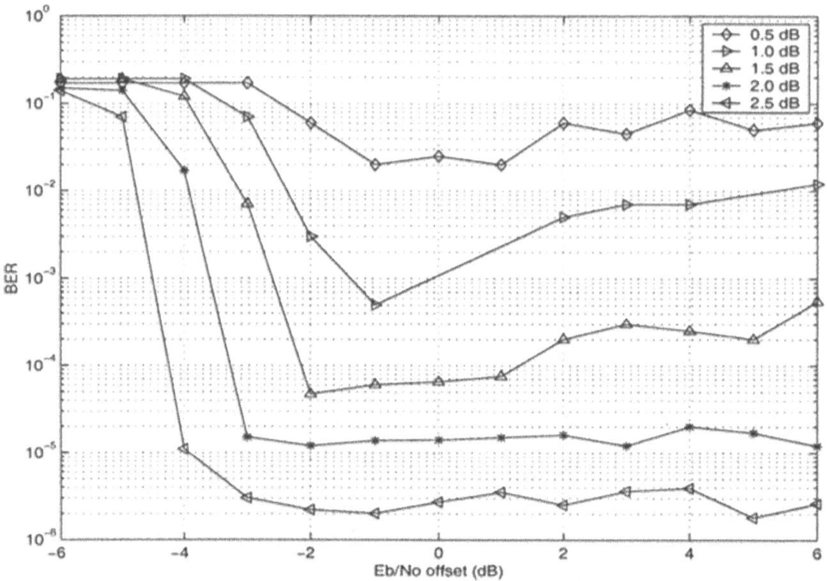

Figure 8.16. Effect of Channel SNR Mismatch on Performance of PTCC

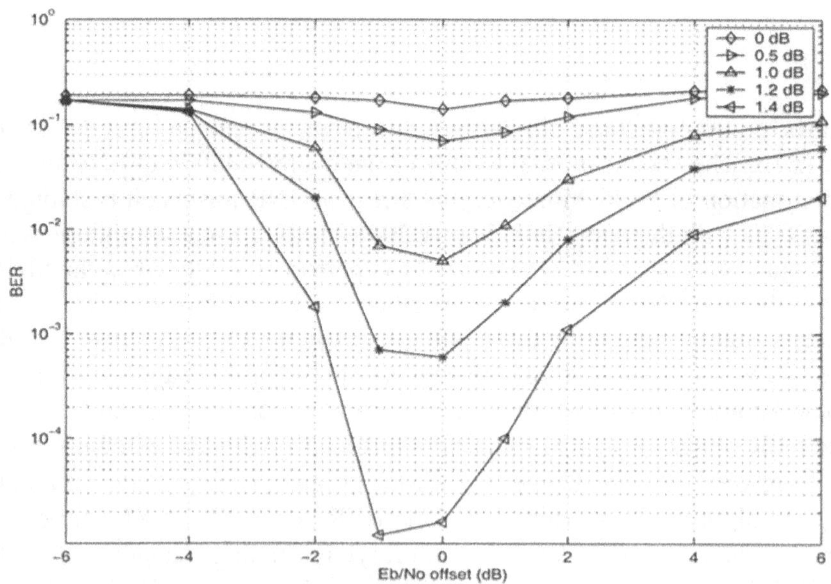

Figure 8.17. Effect of Channel SNR Mismatch on Performance of STCC

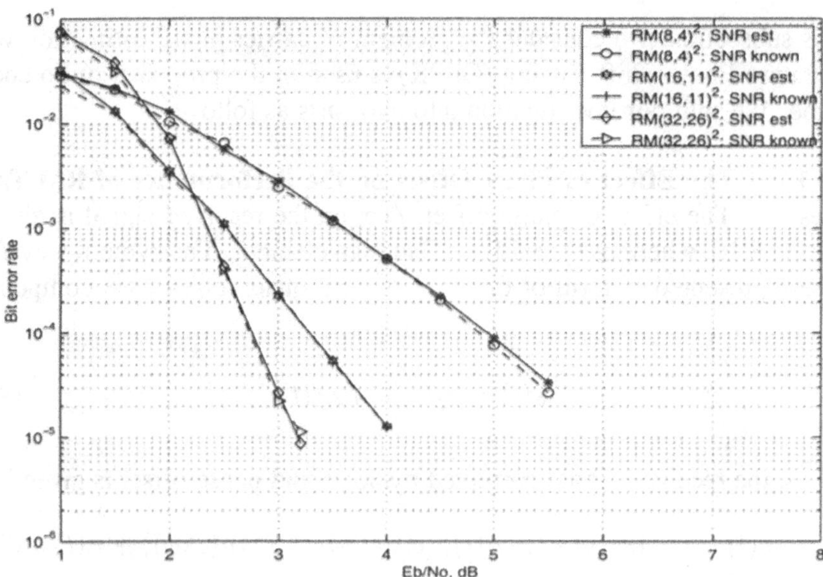

Figure 8.18. Performance of a RM $(32, 26)^2$-turbo Code with and without Variance Estimation on a Gaussian Channel

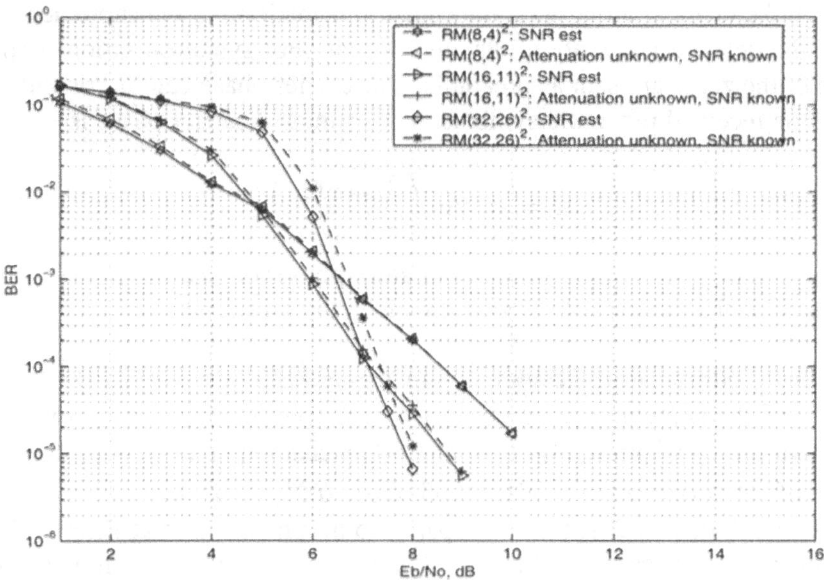

Figure 8.19. Performance of a RM $(32, 26)^2$-turbo Code with and without Variance Estimation on a Rayleigh Fading Channel

8.3.3 Carrier Phase Recovery

In this subsection, we consider the problem of carrier phase estimation when Quadrature Phase-Shift Keying (QPSK) is used in the proposed turbo coding scheme. We split our investigation into two parts as follows:

8.3.3.1 The Effect of Phase Offset on the Performance of RM Turbo Codes. The effect of phase offset, $\Delta\phi$, on the received signal is given in Equation 8.10, where $\alpha(t) = 1$ for a Gaussian channel. The symbol waveform can be represented in term of the in-phase and quadrature-phase components as,

$$s(t) = s_I(t) + j\, s_Q(t) \tag{8.14}$$

Thus, the received signal corrupted by noise and phase offset is given by,

$$\begin{aligned} r(t) &= (s_I(t) + j\, s_Q(t))\,(cos(\Delta\phi) + j\, sin(\Delta\phi)) + n(t) \qquad (8.15)\\ &= (s_I(t)cos(\Delta\phi) - s_Q(t)sin(\Delta\phi)) \\ &\quad +j\ (s_Q(t)cos(\Delta\phi) + s_I(t)sin(\Delta\phi)) + n(t) \end{aligned}$$

8.3.3.2 The Effect of Preamble Size on the Performance of RM Turbo Codes. One way to recover the carrier phase is to send uncoded preamble bits through the channel along with the coded information. In the shortened turbo code, the deleted information bits are not sent. For synchronization purpose, some of the zeros are sent as preamble. The carrier phase can be computed by using the received preamble symbols. The estimated carrier phase is

$$\widehat{\phi} = \arctan\left(\frac{\sum_{i=0}^{V} r_{Ii}}{\sum_{i=0}^{V} r_{Qi}}\right) \tag{8.16}$$

where r_{Ii} and r_{Qi} are in-phase and quadrature-phase components of the received signal. V is the number of QPSK symbols in the preamble.

8.3.3.3 Simulation Results. In Figure 8.20, we give the BER versus E_b/N_o of the shortened turbo code of case C (see Section 6.4.6.1) with different variances of phase offset on an AWGN channel. The $\frac{E_b}{N_o}$ loss due to the phase offset variances of 0.001, 0.005, 0.01, 0.02, 0.04, 0.06, 0.1 $(rad)^2$ are 0, 0.1, 0.2, 0.4, 1.4, 2.4, 5 dB compared to no phase offset at BER of 10^{-4}. However, at variance of 0.22 $(rad)^2$, there is an error floor which means very little improvement in terms of performance can be obtained by increasing the E_b/N_o

In Figure 8.21, we present the BER curve showing the effect of preamble size with $\Delta\phi \sim U\ [-\pi, \pi]$. The results show that with 50 preamble symbols

(100 bits), the effect of phase offset is completely removed. A preamble length of 25 results in 0.25 dB degradation.

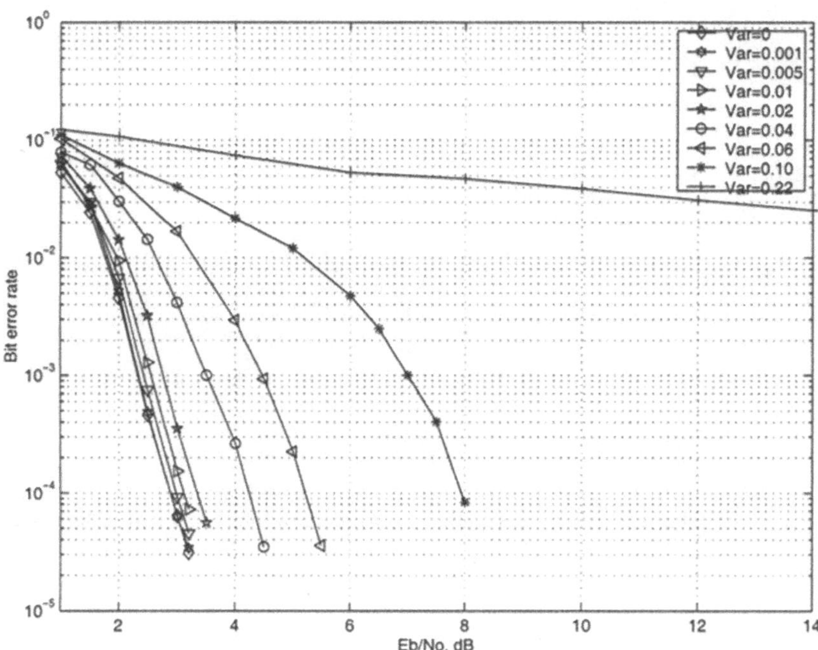

Figure 8.20. Effect of Phase Offset on the Performance of Shortened RM-turbo Code Case C.

8.4. Hardware Implementation of Turbo Codes

In the report of DVB RCS-272, the principle of the turbo decoding for DVB-RCS standard is discussed. Figure 8.22 gives the generic processing engine of the turbo decoder. This engine is built around two SISO modules. The SISO modules are identical in structure, however, as inputs, one receives data in the natural order and the other in the interleaved order. The output of one SISO, after proper scaling and after reordering, is used by its dual SISO in the next step.

The implementation trade-offs are addressed in this report. A prologue of 32 trellis steps is used to find out the right circular state. 4-bit quantization can get good performance for the complexity/performance compromise. Based on 4-bit quantization, three types of memories in a turbo decoder are discussed, including the input buffer memory, the metrics memory and the extrinsic information memory. Table 8.4 shows the typical silicon requirements for two implementations, one on the FPGA and another on an ASIC qualified for space application.

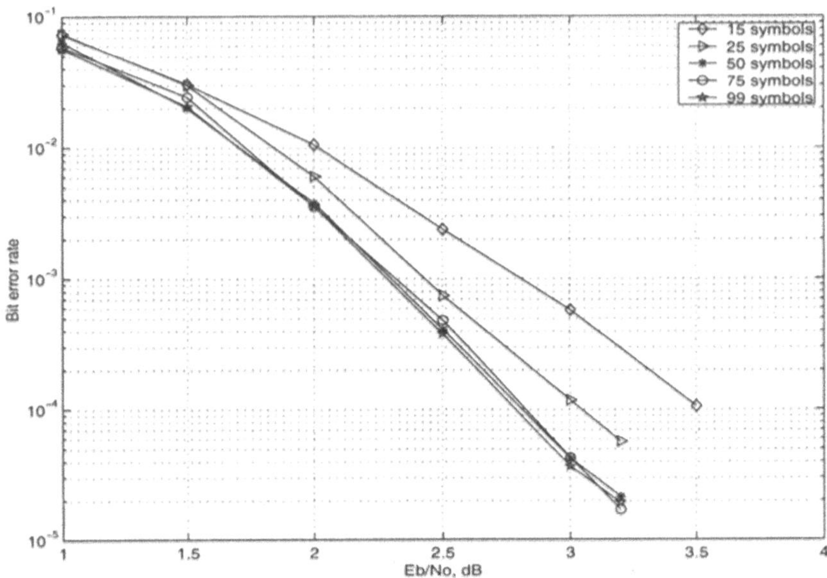

Figure 8.21. Effect of Preamble Sizes on the Performance of Shortened RM-turbo Code Case C.

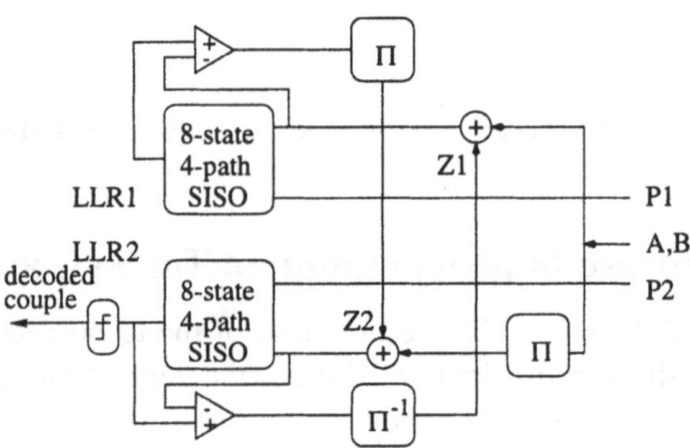

Figure 8.22. The Principle of the Turbo Decoding for DVB-RCS Standard

Signal processing functions such as the decoding of turbo codes can be implemented using general purpose Digital Signal Processors (DSPs). For example, the decoder for a 3GPP turbo decoder can be implemented using Texas Instruments'TMS320c6201. The implementation on this DSP can decode upto an information rate of 440 Kbps for 4 decoding iterations [106].

Tables 8.5, 8.6, 8.7 and 8.8 show some examples of the implementation of turbo codes on different DSP families. Tables 8.5, 8.6 show the results for a 16-state $K = 5$. This code uses the so called TURBO4 [161] feedback

Parameters	*FPGA/CPLD*	*ASIC (space qualified)*
Input quantization	4 bits	4 bits
Number of inputs and guaranteed user rate(s)	one input of 4 Mbits (6 iterations)	one 6.3 Mbits input or three asynchronous inputs of 2.1 Mbits each (12 iterations)
Number of input buffer blocks	2	$6^{(*)}$
Number of processing engines	1	$3^{(*)}$
FPGA/ASIC	ALTERA APEX20KE200 (200Kgates)	TEMIC MG2RT
Typical use of silicon	Memory: about 100% RAM cells Algorithm: about 6000 Logic Elements	Input memory < $6 \times 16.5 = 100Kgates$ Turbo decoders (including other memories) < $3 \times 75 = 225Kgates$

Table 8.4. Typical Silicon Requirements

(*) This architecture is used due to the particular requirement of the target application, i.e., that the ASIC must be capable of processing, in parallel, three asynchronous bit streams each one-third of the 6.3 Mbits.

and feed-forward connection polynomials $(23)_{octal}$ and $(35)_{octal}$, respectively. Given the fact that the decoding complexity is proportional to the number of states, the throughput presented in these two tables should be roughly doubled if the same design is applied to the 3GPP decoder. According to the designers' estimate, a speed up factor of around 1.7 should be expected[114].

Table 8.5 shows the performance of theTURB04 decoder implemented on ADSP-2181 by K. Gracie *et al.* of Communications Research Center (CRC) [112]. ADSP-2181 is a fixed-point 16-bit, 40 MIPS processor from Analog Devices. Table 8.6 shows the performance of the decoder implemented by the same team on ADSP-2106x SHARC DSP from Analog Devices [113]. Apart from the change in the DSP, the increased throughput of the implementation on SHARC is due to:

- The metric combining algorithm is different and much better suited to the Analog Devices DSP architecture[115]

- An early stopping criterion that significantly increases throughput as the SNR increases is also available [113]. With a block size of 512 information bits, operating at $BER = 4 \times 10^{-5}$, the SHARC decoder using early stopping delivers 8 iteration performance at a throughput of 75.1 kbps. Approaching $BER = 10^{-6}$, the decoder delivers 8 iteration performance at over 80 kbps [114].

- Sub-block processing [116] is used to greatly increase the range of block sizes that can be supported. Without sub-block processing, the SHARC decoder using only internal memory is able to support a maximum block size of 650 information bits at rate 1/2 or higher; with sub-block processing, block sizes of thousands of information bits are easily achieved. For example, assuming rate 1/2 or higher, using ADSP-21061 (0.5 Mbits of internal memory), the maximum block size is roughly 3000 information bits, while using an ADSP-21060 (2.0 Mbits of internal memory) yields a maximum block size of roughly 18,000 information bits [114].

The results shown in Tables 8.5, 8.6 are obtained using a simple *relative prime* (RP) interleaver and and zero flushing of the first trellis.

Table 8.7 shows the performance obtained by Jason P. Woodard [117] for the 3GPP turbo code on Texas Instruments' 200 MHz. DSP, TMS320C62x. The fastest implementation on a general purpose processor is shown in Table 8.8. This implementation is done on the 160 MHz. TMS320C6201 DSP and can go up to 440 Kbps [106].

Cons trait Length	*Decoding Speed*	*Data Block Size*	E_b/N_0 for BER=10^{-4}
$K = 5$	16.8 (4 ite.)	$N = 64 - 512$	1.4dB (r = 1/3, $N = 512$, ite=4)
			2.0dB (r = 1/2, $N = 512$, ite=4)
	8.5 (8 ite.)		1.3dB (r = 1/3, $N = 512$, ite=8)

Table 8.5. The Implementation of TURB04 Decoder on ADSP-2181.

Constraint Length	*Decoding Speed*	*Data Block Size*	E_b/N_0 for BER=10^{-4}
$K = 5$	47.1 (4 ite.)	$N = 64 - 1024$	1.75dB (r = 1/2, $N = 1024$, ite=4)
	22.8 (8 ite.)		1.6dB (r = 1/2, $N = 1024$, ite=8)

Table 8.6. The Implementation of TURB04 Decoder on ADSP-2106x SHARC

Constraint Length	*Decoding Speed*	*Data Block Size*	E_b/N_0 for BER=10^{-4}
$K = 4$	144 (4 ite.)	$N = 1440$	1.3dB (r = 1/3, $N = 1440$, ite=4)

Table 8.7. The Implementation of 3GPP Decoder on TMS320C62x

Another possibility is to use DSPs designed with the specific application in mind. For example, in the case of the 3GPP turbo decoder, one can use the new

Constraint Length	*Decoding Speed*	*Data Block Size*	E_b/N_0 for BER=10^{-4}
$K = 4$ (scalable)	up to 440 (4 ite.)	N up to 1440 (scalable)	$K = 4$, St = 8: 1.20dB (r = 1/3, N = 1440, ite=4) 1.80dB (r = 1/2, N = 1440, ite=4)

Table 8.8. The Implementation of 3GPP Decoder on TMS320C6201

TM320c6416 DSP designed by Texas Instruments. This device is the highest-performance DSP CPU so far developed by TI. This DSP has two embedded coprocessors: Viterbi coprocessor and Turbo Decoder coprocessor, which has targeted specifically 3GPP Base Station hardware market. TMS320C6416 is capable of decoding up to 12Mbps (6 iterations). The turbo decoder of this DSP CPU is exactly designed for 3GPP turbo code and has some parameters to be set for decoding any type of 3GPP code with any data block size, stop criterion, code rate, etc.

8.5. Summary

In this chapter, the effect of input data quantization for turbo codes using double-binary CRSC component codes and BTCs using RM-turbo codes, were presented. Uniform quantization was used for TCCs and BTCs. For TCCs, adaptive-step size input data quantization was also discussed. In the case of double-binary CRSC code, the performance of the adaptive step size is better than that of the fixed step size. In addition, the choice of step size depends on the code rate. It was shown that 4-bit input quantization is a good tradeoff between the performance and decoding complexity for both TCCs and BTCs. Furthermore, the effect of correction term was also presented. This term is used to improve the performance when Max-log-MAP algorithm is used instead of log-MAP algorithm. The simulation results show that there is no significant degradation even when a simple 1-bit look-up table is used.

We also presented the effect of channel impairments, including channel SNR mismatch and the phase offset, on the performance of turbo codes. The effect of preamble size used to recover the carrier phase was also presented. The results showed that parallel concatenated codes are more sensitive to the underestimation of the SNR than to the overestimation of the SNR. The tolerance of SNR mismatch was in the range of -2 to 6 dB for RM $(32, 26)^2$ turbo code. Shorter codes were more tolerant. Serial concatenated convolutional code is sensitive to both overestimation and underestimation of the SNR. Moreover, the effect of phase offset on RM $(32, 26)^2$ turbo code was discussed and it was concluded that a small phase offset (variance less than 0.02) is bearable. However, beyond this, carrier phase offset should be compensated. We showed that a preamble size of 50 symbols for QPSK modulation scheme was enough to recover the carrier phase completely.

Hardware implementation of turbo codes was also discussed and several examples of hardware development for FPGA, ASIC and DSP were presented.

Chapter 9

LOW DENSITY PARITY CHECK CODES

In 1962, R.G.Gallager [162] introduced a class of error correcting codes called *Low-Density Parity-Check* (LDPC) codes. These codes have parity check matrices that are sparse, *i.e.*, contain mostly 0's and have only a few 1's . Although the sparseness of the parity check matrix results in low decoding complexity, still the decoding complexity was high enough to make the implementation of the LDPC codes infeasible until recently. It is interesting to note that the iterative decoding procedure proposed by Gallager [162] is practically the same as the message passing schemes used for decoding of the turbo and turbo-like codes today. In spite of all this, apart from a few references [164] [165] [163] to Gallager's work, overall the subject remained unknown to the information theory community. It was only after the discovery of turbo codes in 1993 [6] that interest in Low-Density Parity-Check codes was rekindled and LDPC codes were re-discovered independently by MacKay and Neal [167] and Wiberg [166]. In the past few years, there has been a considerable amount of research work on LDPC codes [168], [171], [14], [176], [174], [179], [178] and [180].

9.1. Gallager Codes: Regular Binary LDPC Codes

Coding for error correction is one of the many tools available for achieving reliable data transmission in communication systems [162]. Shannon showed that for any channel with a defined capacity, there exist coding schemes that, when decoded with an optimal decoder, achieve arbitrarily small error probability for all transmission rates below capacity. A fundamental problem of information theory is to make practical codes whose performance approaches the Shannon limit. The diagram of general error-correcting communication system is depicted in Figure 9.1.

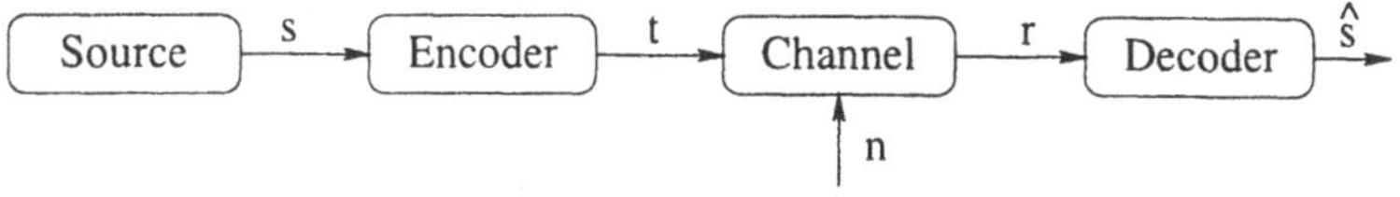

Figure 9.1. Diagram of General Error-correcting Communication System.

The aim of an error-correction coding scheme is to encode the information sequence in such a way that the distribution of the encoded symbols is very close to the probability distribution that maximizes the mutual information between the input and the output of the channel.

By doing this, an error-correcting code minimizes the probability of residual errors after decoding, while introducing as little redundancy as possible during encoding. The codes that Shannon used in his proof were random block codes, which are introduced in the next section.

9.2. Random Block Codes

Consider a channel with the input alphabet $\mathcal{X}$ and the output alphabet $\mathcal{Y}$. We make the following definitions:

An (n, k) block code is a mapping from $\{0,1\}^k$ to $\mathcal{X}^n$. A binary input message $\mathbf{x}$ of length k is mapped to a codeword $\mathbf{t} \in \mathcal{X}^n$ of length n.

The rate of communication is $R = k/n$, *i.e.*, k bits of information are sent in n channel uses.

A decoder is a mapping from $\mathcal{Y}^n$ to $\{0,1\}^k$. Received channel outputs $\mathbf{r} \in \mathcal{Y}^n$ are mapped to the information sequence $\hat{\mathbf{x}} \in \mathcal{X}^k$.

The probability of block error of a code, given a distribution over input messages and a channel model, is:

$$p_B = \sum_{\mathbf{x}} P(\mathbf{x}) P(\hat{\mathbf{x}} \neq \mathbf{x} | \mathbf{x}) \tag{9.1}$$

The optimal decoder is the decoder that minimizes p_B.

Figure 9.2 summarizes the operation of an (n, k) block code.

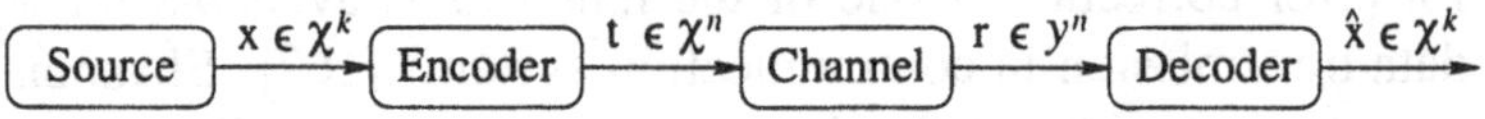

Figure 9.2. Outline of (n, k) Block Code.

According to Shannon's channel coding theorem: for any rate $R < C$, and any $\varepsilon > 0$, there is some N such that for any $n > N$, there are (n, k) codes with $R = k/n$ that ensure that the probability of error does not exceed ε. Also, Shannon's proof of the channel coding theorem indicates that for a large class of channels, almost all randomly selected long codes are "good" in the above sense.

The abundance of "good" codes, however, does not translate itself into the ease of finding easily decodable codes. Shannon relates this problem to the difficulty of "giving an explicit construction for a good approximation to a random sequence" [1].

9.2.1 Generator Matrix

A linear block code of block length n and rate k/n can be described by a generator matrix $\mathbf{G}$ of dimension $k \times n$ that describes the mapping from source words $\mathbf{s}$ to codewords $\mathbf{t} := \mathbf{G}^T\mathbf{s}$ (where the vectors, $\mathbf{s}$ and $\mathbf{t}$ are column vectors). It is common to consider $\mathbf{G}$ in systematic form, $\mathbf{G} \equiv [\mathbf{I}_k|\mathbf{P}]$ so that the first k transmitted symbols are the source symbols. The notation $[\mathbf{A}|\mathbf{B}]$ indicates the concatenation of matrix $\mathbf{A}$ with matrix $\mathbf{B}$; $\mathbf{I}_k$ represents the $k \times k$ identity matrix. The remaining $m = n - k$ symbols are the parity-checks.

9.2.2 Parity Check Matrix

A linear block code is also described by a parity check matrix $\mathbf{H}$ of dimension $m \times n$, where $m = n - k$. If the corresponding generator matrix is written in systematic form as above, then $\mathbf{H}$ has the form $[-\mathbf{P}^T|\mathbf{I}_m]$. Note that for codes over finite fields $GF(2^p)$, $-\mathbf{P} \equiv \mathbf{P}$. Each row of the parity-check matrix describes a linear constraint satisfied by all codewords. $\mathbf{H}\mathbf{G}^T = 0$ and hence the parity-check matrix can be used to detect errors in the received vector:

$$\mathbf{Hr} = \mathbf{H}(\mathbf{t} + \mathbf{e}) = \mathbf{HG}^T\mathbf{s} + \mathbf{He} = \mathbf{He} := \mathbf{z}$$

where $\mathbf{e}$ is the error vector and $\mathbf{z}$ is the syndrome vector. If the syndrome vector is null, we assume that there has been no error. Otherwise, the decoding problem is to find the most likely error vector $\hat{\mathbf{e}}$ that explains the observed syndrome given the assumed properties of the channel. The operation of the linear error-correcting codes is summarized in Figure 9.3.

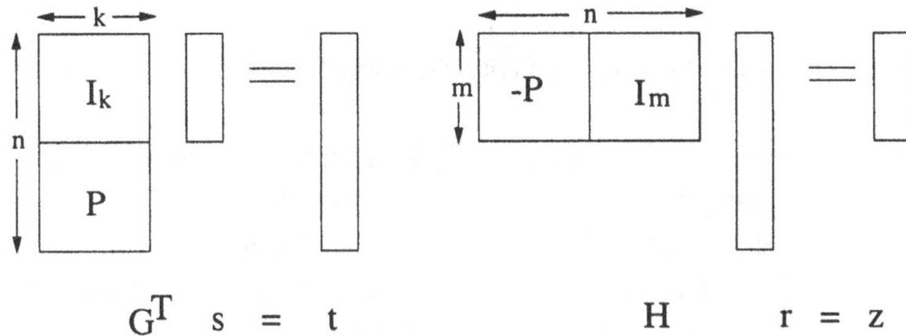

Figure 9.3. Linear Error-correcting Codes: $\mathbf{G}$ maps a message $\mathbf{s}$ to a transmitted codeword $\mathbf{t}$. During transmission the channel adds noise resulting in error $\mathbf{e}$. $\mathbf{H}$ maps received message $\mathbf{r} := \mathbf{t} + \mathbf{e}$ to syndrome $\mathbf{z}$.

9.3. Regular Binary LDPC Codes: Original Gallager Codes

Low-density parity-check codes are defined in terms of a sparse parity-check matrix $\mathbf{H}$ that consists almost entirely of zeroes. Gallager defined (n, p, q)

LDPC codes to have a block length n and a sparse parity-check matrix with exactly p 1's per column and q 1's per row, where $p \geq 3$ and $q > p$. Figure 9.4 shows a ($n = 20, p = 3, q = 4$) code constructed by Gallager [162]. In this code , every codeword bit participates in exactly p parity-check equations and every such check equation involves exactly q codeword bits. If all the rows are linearly independent then the rate of the code is $(q - p)/q$, otherwise the rate is $(n - m)/n$, where m is the dimension of the row space of $\mathbf{H}$.

1	1	1	1	0	0	0	0	0	0	0	0	0	0	0	0	0	0	0	0
0	0	0	0	1	1	1	1	0	0	0	0	0	0	0	0	0	0	0	0
0	0	0	0	0	0	0	0	1	1	1	1	0	0	0	0	0	0	0	0
0	0	0	0	0	0	0	0	0	0	0	0	1	1	1	1	0	0	0	0
0	0	0	0	0	0	0	0	0	0	0	0	0	0	0	0	1	1	1	1
1	0	0	0	1	0	0	0	1	0	0	0	1	0	0	0	0	0	0	0
0	1	0	0	0	1	0	0	0	1	0	0	0	0	0	0	1	0	0	0
0	0	1	0	0	0	1	0	0	0	0	0	0	1	0	0	0	1	0	0
0	0	0	1	0	0	0	0	0	0	1	0	0	0	1	0	0	0	1	0
0	0	0	0	0	0	0	1	0	0	0	1	0	0	0	1	0	0	0	1
1	0	0	0	0	1	0	0	0	0	0	1	0	0	0	0	0	1	0	0
0	1	0	0	0	0	1	0	0	0	1	0	0	0	0	1	0	0	0	0
0	0	1	0	0	0	0	1	0	0	0	0	1	0	0	0	0	0	1	0
0	0	0	1	0	0	0	0	1	0	0	0	0	1	0	0	1	0	0	0
0	0	0	0	1	0	0	0	0	1	0	0	0	0	1	0	0	0	0	1

Figure 9.4. Example of a Low-density Parity-check Matrix for a (20, 3, 4) LDPC Code

In Gallager's construction of Figure 9.4, the matrix is divided into p submatrices, each containing a single 1 in each column. The first of these submatrices contains all its 1's in descending order; *i.e.*, the i'th row contains all its 1's in column's $(i - 1)q + 1$ to iq. The lower two sections of the matrix are column permutations of the upper section. By considering the ensemble of all matrices formed by such column permutations, Gallager proved several important results. These include the fact that the error probability of the optimum decoder decreases exponentially for sufficiently low noise and sufficiently long block length, for fixed p. Also, the typical minimum distance increases linearly with the block length.

9.3.1 Construction of Regular Gallager Codes

One of the attractions of the LDPC codes is their simple description in terms of a random sparse parity-check matrix, making it easy to construct for any rate.

Many good codes can be built by specifying a fixed weight for each row and each column, and constructing at random subject to those constraints. However, the best LDPC codes use further design criteria. Here is the basic constraints of Gallager code construction.

- The parity-check matrix has a fixed column weight p and a fixed row weight q.
- The parity-check matrix is divided into p submatrices, each containing a single 1 in each column.
- Without loss of generality, the first submatrix is constructed in some predetermined manner.
- The subsequent submatrices are random column permutations of the first submatrix.

Since $\mathbf{H}$ is not in systematic form, Gaussian elimination using row operations and reordering of columns needs to be performed to derive a parity-check matrix $\mathbf{H}' = [\mathbf{P}^T|\mathbf{I}_m]$. Then the original $\mathbf{H}$ has to be redefined to include the column reordering as per the Gaussian elimination. The corresponding generator matrix is then $\mathbf{G} = [\mathbf{I}_k|\mathbf{P}]$.

$\mathbf{G}$ is not in general sparse, so the encoding complexity is $O(n^2)$ per block. However, with simple modifications of the structure of $\mathbf{H}$, the encoding complexity can be reduced significantly [174].

9.4. Decoding

There are two decoding schemes used to achieve a reasonable balance between the complexity and the probability of decoding error. The first is particularly simple but is applicable only to the BSC at rates far below the channel capacity. The second scheme, which decodes directly from the *a posteriori* probabilities at the channel output, assumes that the code words from an (n, p, q) code are used with equal probability on an arbitrary binary-input channel.

9.4.1 Introduction of Gallager's Decoding

In the first decoding scheme, the decoder computes all the parity checks and then changes any digit that is contained in more than some fixed number of unsatisfied parity-check equations. Using these new values, the parity checks are recomputed, and the process is repeated until the parity checks are all satisfied. If the parity-check sets are small, the decoding complexity is reasonable, since most of the parity-check sets will contain either one transmission error or no transmission errors. Thus when most of the parity-check equations checking on a digit are unsatisfied, there is a strong indication that the digit given is in error. Suppose that a transmission error has occurred in the first digit of the code in Figure 9.4. Then the parity checks 1, 6, and 11 would be violated, and all three parity-check equations checking digit 1 would be violated. On the other hand, at most, one of the three equations checking on any other digit in the block would be violated.

The second decoding scheme, called probabilistic decoding, is an iterative decoding regarding *a posteriori* probabilities via the parity-check set tree. The most significant feature of this decoding scheme is that the computation per digit per iteration is independent of the block length. Furthermore it can be shown that the average number of iterations required to decode is bounded by a quantity proportional to the log of the log of the block length. The weak bound on the probability of error was derived in Gallager's paper [162].

In Figure 9.1, the channel adds noise to the vector $\mathbf{t}$ with the resulting received signal $\mathbf{r}$ being given by

$$\mathbf{r} := (\mathbf{G}^T\mathbf{s} + \mathbf{e}) \; mod \; 2 \tag{9.2}$$

The decoder's task is to infer $\mathbf{s}$ given the received message $\mathbf{r}$, and the assumed noise properties of the channel. The optimal decoder returns the message $\mathbf{s}$ that maximizes the a posteriori probability

$$P(\mathbf{s}|\mathbf{r}, \mathbf{G}) = \frac{P(\mathbf{r}|\mathbf{s}, \mathbf{G})P(\mathbf{s})}{P(\mathbf{r}|\mathbf{G})} \tag{9.3}$$

It is often not practical to implement the optimal decoder. Indeed, the general decoding problem is known to be NP-complete [169]. For generalized Gallager's constructions, the decoding procedure using bipartite graphs is introduced as follows.

9.4.2 Syndrome Decoding Based on Tanner's Graph

For syndrome decoding, the most probable vector $\mathbf{x}$ (according to the channel model) has to be found, which explains the observed syndrome vector $\mathbf{H}\mathbf{x} = \mathbf{z}$. The vector $\mathbf{x}$ is then our estimate of the error vector. The components of $\mathbf{x}$ are the noise symbols. The exact decoding problem is known to be NP-complete even when the column weight is fixed to be 3, therefore, an approximate algorithm must be used. Here we introduce the details of the decoding procedure described in [174].

The iterative probabilistic decoding algorithm is known as a sum/product [166] or belief propagation [177] algorithm. At each step, we estimate the posterior probability of the value of each noise symbol, given the received signal and the channel properties. The process is best viewed as a message passing algorithm on the bipartite graph defined by $\mathbf{H}$ in which we have two sets of nodes: the nodes representing the noise symbols, and the nodes representing the check symbols (See Figure 9.5). Nodes z_i and x_j are connected if the corresponding matrix entry H_{ij} is non-zero. The directed edges show the causal relationships: the state of a check node is determined by the state of the noise nodes to which it is connected. We refer to the neighbors of a noise node x_j as its *children* and to the neighbors of a check node as its *parents*.

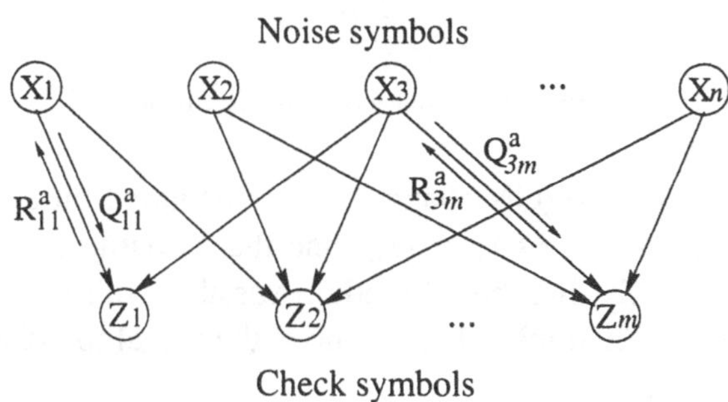

Figure 9.5. Message Passing on the Bipartite Graph Representing a Parity-check Matrix

At each step of the decoding algorithm each noise node x_j sends messages Q^a_{ij} to each child z_i which are supposed to approximate the node's belief that it is in state a (value of 0 or 1 in the binary case), given messages received from all its other children. Also, each check z_i sends messages R^a_{ij} to each parent x_j approximating the probability of check i being satisfied if the parent is assumed to be in state a, taking into account messages received from all its other parents. After each step we examine the messages and produce a tentative decoding. The decoding algorithm consists of iteratively updating these messages until the tentative decoding satisfies the observed syndrome vector (declare a success) or a preset maximum number of iterations is reached (declare a failure). The maximum number of iterations may be set to perhaps ten times the typical number, improving the success rate while imposing little overhead on the average decoding time. Although it is in principle possible for the decoder to converge to the wrong noise vector, this is not observed in practice. That is, (empirically) all decoding errors are detected.

If the underlying graph has a tree structure, the algorithm is known to converge to the true posterior distribution after a number of iterations equal to the diameter of the tree. The problem is that there are many cycles in the graph and occasionally the algorithm fails to converge at all. One should take care to avoid short cycles in the graph.

9.4.2.1 Initialization. The algorithm is initialized by setting each message Q^a_{ij} to f^a_j, the a priori probability that the jth noise symbol is a. In the case of a BSC f^1_j would be equal to the crossover probability.

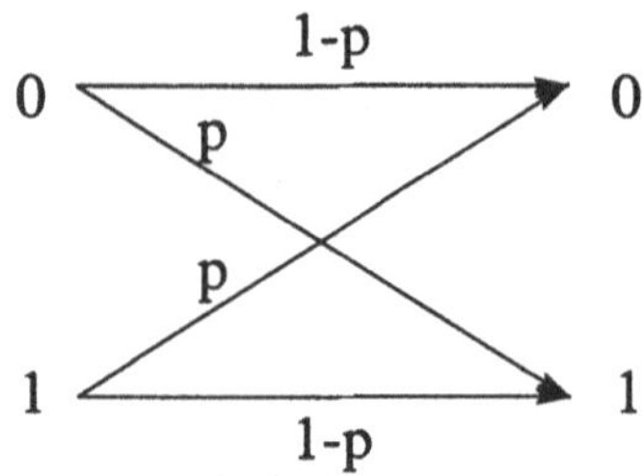

Figure 9.6. Binary Symmetric Channel

For the binary-input AWGN channel, the transmitted bit $t \in (0, 1)$ map to the transmitted signal $x \in (+x_0,\ -x_0)$ and the output is $y = x + v$ where v is a zero mean normally distributed random variable with variance σ^2. We set $\sigma = 1$ and let the signal amplitude x_0 control the signal to noise ratio.

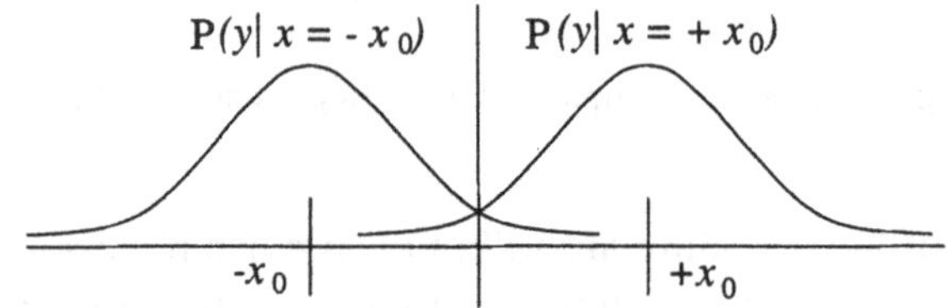

Figure 9.7. Gaussian Channel

We declare the received bit $r = 1$ if $y > 0$ and $r = 0$ if $y < 0$ where $r = t + e \ mod \ 2$. Define

$$\ln \frac{P(e=0|y)}{P(e=1|y)} = \ln \frac{\frac{P(e=0)P(y|e=0)}{P(y)}}{\frac{P(e=1)P(y|e=1)}{P(y)}}$$

$$= \ln \frac{P(y|e=0)}{P(y|e=1)} \tag{9.4}$$

If $y > 0$,

$$\ln \frac{P(y|e=0)}{P(y|e=1)} = \ln \frac{P(y|x=x_0)}{P(y|x=-x_0)}$$

$$= \ln \frac{\frac{1}{\sqrt{2\pi}\sigma} \exp(-\frac{(y-x_0)^2}{2\sigma^2})}{\frac{1}{\sqrt{2\pi}\sigma} \exp(-\frac{(y+x_0)^2}{2\sigma^2})}$$

$$= 2x_0 y$$

If $y < 0$,

$$\ln \frac{P(y|e=0)}{P(y|e=1)} = \ln \frac{P(y|x=-x_0)}{P(y|x=x_0)}$$

$$= \ln \frac{\frac{1}{\sqrt{2\pi}\sigma} \exp(-\frac{(y+x_0)^2}{2\sigma^2})}{\frac{1}{\sqrt{2\pi}\sigma} \exp(-\frac{(y-x_0)^2}{2\sigma^2})}$$
$$= -2x_0 y$$

then

$$\frac{P(y|e=0)}{P(y|e=1)} = \exp[2x_0|y|] \tag{9.5}$$

and

$$\frac{P(e=0|y)}{P(e=1|y)} = \exp[2x_0|y|] \tag{9.6}$$

Since $P(e=0|y) = 1 - P(e=1|y)$, the likelihood of this bit being in error is:

$$f^1 := P(e=1|y) = (1 + exp(2x_0 \mid y \mid))^{-1} \tag{9.7}$$

and $f^0 = 1 - f^1$.

9.4.2.2 Updating R^a_{ij}. The messages R^a_{ij} that check i sends to parent j should be the probability of check i being satisfied if the parent was in state a. In the sense it is used here, check i is 'satisfied' if it agrees with the corresponding syndrome symbol z_i. In syndrome decoding, z_i is not necessarily zero. The laws of probability tell us:

$$P(z_i|x_j = a) = \sum_{x:x_j=a} P(z_i|x)P(x|x_j = a) \tag{9.8}$$

Hence, we sum over all configurations $\mathbf{x}$ for which the check is satisfied and the parent is in state a and add up the probability of the configuration (product of associated Q messages). For node z_i we update the outgoing message to node x_j for each value a as follows:

$$R^a_{ij} = \sum_{x:x_j=a} P(z_i|x) \prod_{k\in\aleph(i)\backslash j} Q^{x_k}_{ik} \tag{9.9}$$

where $\aleph(i)$ denotes the set of indices of the parents of the node z_i and $\aleph(i) \setminus j$ denotes the indices of all parents except node j. The probability $P(z_j|x)$ of the check being satisfied is either 0 or 1 for any given configuration $\mathbf{x}$.

R can be calculated efficiently by treating the partial sums of a parity check as the states in a Markov chain, with transition probabilities given by the appropriate Q values. The forward-backward algorithm is used to calculate the forward and backward probabilities

$$F_{ij}(a) = P(\sum_{k<j} H_{ik}x_k = a) \tag{9.10}$$

$$B_{ij}(a) = P(\sum_{k>j} H_{ik}x_k = a) \tag{9.11}$$

according to the probabilities given by the Q messages. The calculation of R_{ij}^a is then straightforward:

$$R_{ij}^a = \sum_{\{s,\, t:\; H_{ij}a+s+t=z_i\}} F_{ij}(s)B_{ij}(t) \tag{9.12}$$

9.4.2.3 Updating Q_{ij}^a. The messages Q_{ij}^a that noise node j sends to check i should be the belief the parent has that it is in state a, based on the information from all other children. Applying Bayes' theorem:

$$P(x_j = a|\{z_i\}_{i\in\Im(j)\backslash i}) = \frac{P(x_j = a)P(\{z_i\}_{i\in\Im(j)\backslash i}|x_j = a)}{P(\{z_i\}_{i\in\Im(j)\backslash i})} \tag{9.13}$$

Treating the symbols of z as independent, we take the product of all the other children's votes for state a, weighted by the prior. For node x_j we update the outgoing message to node z_i for each value a as follows:

$$Q_{ij}^a = \alpha_{ij} f_j^a \prod_{k\in\Im(j)\backslash i} R_{kj}^a \tag{9.14}$$

where $\Im(j)$ denotes the set of indices of the children of node x_j and f_j^a is the prior probability that x_j is in state a. The normalizing constant α_{ij} ensures $\sum_a Q_{ij}^a = 1$. The update may be implemented using a forward-backward algorithm.

9.4.2.4 Tentative Decoding. After updating the Q and R messages we calculate, for each index $j \in \{1, ..., n\}$ and possible states a, the quantity

$$\hat{e}_j = argmax\ f_j^a \prod_{k\in\Im(j)} R_{kj}^a \tag{9.15}$$

The vector $\hat{\mathbf{e}}$ is the tentative error vector. If this satisfies the syndrome equation $H\hat{\mathbf{e}} = \mathbf{z}$ then we terminate the decoding, declaring a success. Otherwise we iterate, updating R and Q again until either decoding is successful or we declare a failure after a fixed number of iterations (for example, 500). Figure 9.8 shows the evolution of the bit error probability as a function of the iteration number [178].

The most significant feature of this decoding scheme is that the computation per digit per iteration is independent of the block length. Furthermore it can be shown that the average number of iterations required to decode is bounded by a quantity proportional to the log of the log of the block length.

9.5. New Developments

Gallager's codes attracted little attention prior to 1995, but there has been a recent surge of interest since their performance was recognized. Davey and

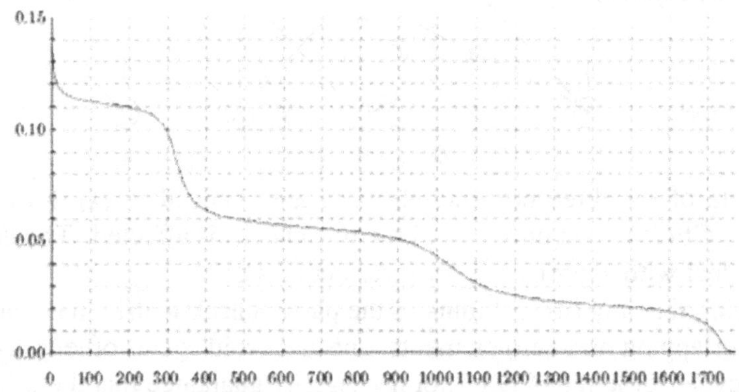

Figure 9.8. Evolution of the Bit Error Probability as a Function of the Iteration Number

MacKay [13] introduced non-binary versions of Gallager's codes. For non-binary versions of the codes, messages are encoded using symbols from a finite field with more than two elements, each parity-check becomes more complex but decoding remains tractable. Although the non-binary codes have an alternative representation as binary codes, the non-binary decoding algorithm is not equivalent to the binary algorithm. These changes can help if the parity-check matrix is constructed carefully.

Luby, Mitzenmacher, Shokrollahi and Spielman [171] introduced parity-check matrices with highly non-uniform column-weight distributions. In 1998, Davey and MacKay [173] presented irregular non-binary codes that outperformed the best known turbo codes. Gallager considered codes whose parity-check matrix had fixed row and column weights (a construction referred to as 'regular'). They relaxed this constraint and produced 'irregular' LDPC codes that have a variety of row and column weights. High weight columns help the decoder to identify some errors quickly, making the remaining errors easier to correct.

9.5.1 MacKay's Constructions

The presense of short cycles in the bipartite graph of the LDPC codes result in the loss performance in the belief propagation decoder. The next figure shows a fragment of a bitpartite graph with short cycles of length 4 indicated by the bold lines.

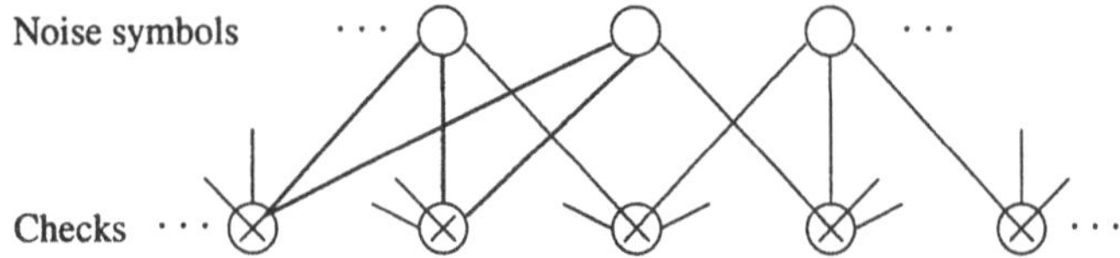

If the state of the three noise symbols x_i are changed to ($x_i + c$) for an arbitrary $c \in$ GF(2^p), then only one of the five checks is affected. The decoder then has to deal with a wrong majority verdict of 4 to 1.

By ensuring that any two columns of the parity-check matrix have no more than one overlapping non-zero elements, one can avoid cycles of length 4 [14]. Having no cycles of length 4 does not necessarily guarantee that the minimum distance is greater than 4 [174]. The example shown in the next figure shows that a case where minimum cycle length is 6 while the minimum distance is 4.

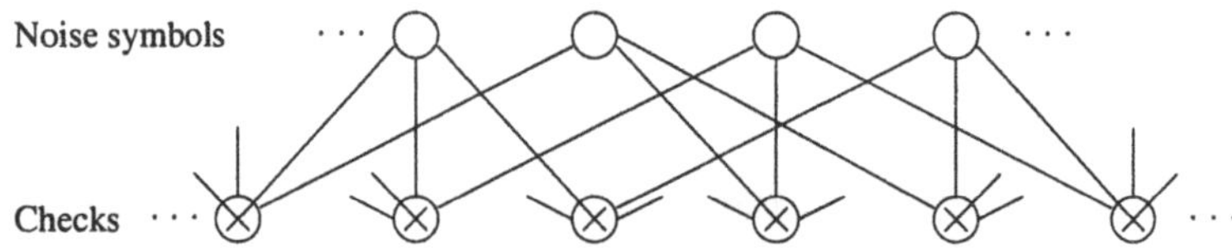

Situations like this, however, are rare since almost for all randomly generated codes, the minimum distance increases linearly with the blocklength. MacKay [14] presented several methods for constructing codes with no cycles of length 4. These methods are listed below:

Construction 1A In this technique, the column weight t is fixed, e.g., t=3 and columns of weight t_c are added to the matrix at random, keeping the row weight, t_r as uniform as possible while avoiding overlap of, more than one between any two columns. This is shown in Figure 9.9 (a).

Construction 2A This construction is similar to 1A. The onle differece is that some (up to m/2) of the columns can have weight 2. These columns are constructed by placing one $\frac{m}{2} \times \frac{m}{2}$ identity matrix on top of another as shown in Figure 9.9 (d).

Constructions 1B, 2B In these construction methods, some of the columns of the 1A or 2A matrices are deleted, in such a way that the bipartite graph of the resulting matrix does not have cycles of length less than l, (e.g. , $l = 6$).

With binary matrices, adding more than $\frac{m}{2} \times \frac{m}{2}$ weight-2 columns results in an increased probability of low weight codewords. With non-binary codes, however, it is possible to add more weight-2 columns [174]. The resulting matrix is called an Ultra-light matix. Two construction techniques for Ultra-light matrices called UL-A and UL-B are given in [174]. These techniques can be considered as a recursive extention of 2A construction.

UL-A After constructing a matrix with $\frac{m}{2}$ weight-2 identity matrices, place two $\frac{m}{4} \times \frac{m}{4}$ indentity matrices next to one of previous $\frac{m}{2} \times \frac{m}{2}$ identity matrices. This process is repeated until m columns of weight-2 have been constructed. This scheme is shown in Figure 9.9 (e).

UL-B construction is similar to UL-A, except that the smaller identity blocks are placed so that each row has weight of at most 2 before the higher weight columns are filled. This scheme is shown in Figure 9.9 (f).

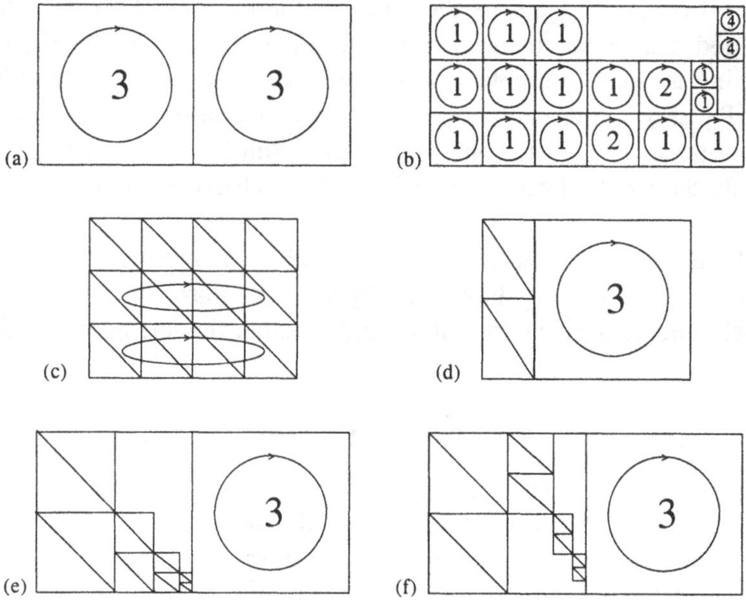

Figure 9.9. Schematic Illustration of Constructions of LDPC Codes. (a) construction 1A for a code with t=3, $t_r = 6$ and rate 1/2; (b) variant of construction IR for a code with rate 1/2; (c) Gallager's construction for a code with rate 1/4; (d) construction 2A for a code with rate 1/3; (e) construction UL-A for a code with rate 15/31; (f) construction UL-B for a code with rate 15/31; (Adapted from diagrams by MacKay [176]).
Notation: an integer in a circle represents a number of permutation matrices superposed on the surrounding square. A diagonal line represents an identity matrix. An ellipse imposed on a block within the matrix represents a random permutation of all the columns in that block.

9.5.2 Irregular Matrices

In the original Gallager codes, all the columns (also all the rows) of the parity check matrix have the same weight. These are called the regular LDPC codes. One can construct codes whose parity check matrix has columns (and rows) with different weights. A method for the construction of the parity check matrix for irregular LDPC codes was proposed by Luby et al. [171]. Here, we present a brief summary of this construction scheme as given in [174]. Readers interested in more detail, may refer to [172].

Construction IR Let λ_i and ρ_i denote the fraction of columns and rows with weight i and, n and m denote the block length and the number of parity checks, respectively. Then the total number of non-zero elements in the parity check matrix is,

$$T = n \sum_i i\lambda_i = m \sum_i i\rho_i$$

The second equality expresses the fact that the number of edges incident to the left nodes is equal to the number of edges incident to the right nodes. Considering a bipartite graph with T 'left nodes' $\{L_1, \ldots L_T\}$ and T 'right nodes' $\{R_1 \ldots R_T\}$. For each columns of weight j in our matrix, label j left (message) nodes with that column's index. Similarly, label i right (check) nodes with the index of each reow of weight i. Then connect each node L_i to R_i.

Finally, the parity check matrix is obtained by permuting the labels of the right nodes while avoiding duplicate edges, *i.e.*, to make sure that the ρ_i right labels beloging to a given row of weight i match the left nodes of different columns.

9.6. Performance Analysis of LDPC Codes

The performance of error-correcting codes was bounded by Shannon in 1948 [1]. However, until the arrival of turbo codes in 1993 [6], practical coding schemes for most non-trivial channels fell far short of the Shannon limit. Turbo codes marked the beginning of near Shannon limit performance for the additive white Gaussian noise channel. Two years later MacKay and Neal rediscovered Gallager's long neglected low-density parity-check codes and showed that, despite their simple description, they too have excellent performance.

9.6.1 Comparison of Empirical Results

Figure 9.10 [174] presents the performance of different LDPC codes and turbo codes showing that they can match and sometimes exceed the performance of turbo codes. All codes shown have rate 1/4. The aim is to achieve the lowest bit error rate for a given signal to noise ratio. That is, the best codes lie towards the bottom-left of the figure.

On the right is a good regular binary LDPC code, as reported by MacKay [14]. Such codes were introduced by Gallager in 1962 but their quality was not recognized until the computing power allowed sufficiently long block length versions to be implemented. The curve labeled '4Galileo' shows the performance of a concatenated code developed at NASA's Jet Propulsion Laboratory based on a constraint length 15, rate 1/4 convolutional code. This code was developed for deep space communication and requires an extremely computer intensive decoder. Until it was eclipsed by turbo codes, it represented the state of the art in error-correction.

Luby *et. al* first investigated irregular constructions of LDPC codes and reported the results labeled 'Luby'. Their methods for choosing matrix parameters are not directly applicable to non-binary codes so alternative construction methods are developed in [174]. The binary irregular code labeled 'Irreg GF(2)' was constructed using the alternative methods for finding construction parameters. Although the block length is just 1/4 the length of the 'Luby' code, the performance is considerably better.

Regular LDPC codes defined over non-binary fields can outperform the binary irregular codes, as shown by the code labeled 'Reg GF(16)', a regular code defined over the finite field with 16 elements.

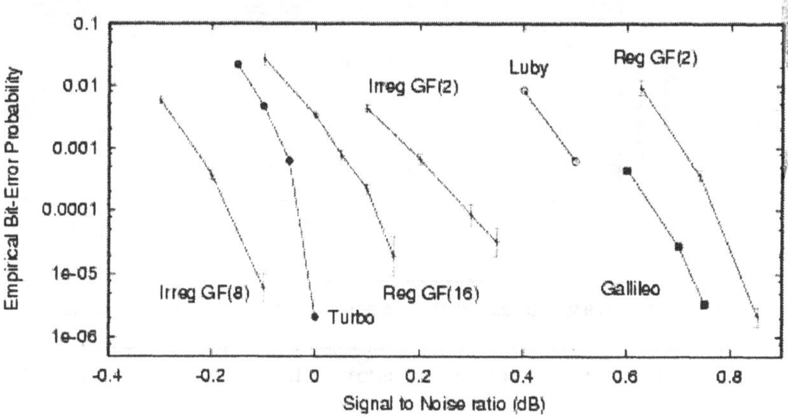

Figure 9.10. Comparison of Empirical Results for Rate 1/4 Improved Low-density Parity-check Codes over the Gaussian Channel. The Shannon limit is at about -0.79dB. From left to right:
Irregular LDPC over GF(8), block length 48000 bits;
JPL turbo code [170], block length 65536 bits;
Regular LDPC over GF(16), block length 24448 bits;
Irregular binary LDPC, block length 16000 bits;
Luby et al. [171] irregular binary LDPC, block length 64000 bits;
JPL concatenated code based on constraint length 15, rate 1/4 convolutional code;
Regular binary LDPC, block length 40000 bits [14].

The code 'Irreg GF(8)' was constructed by combining both modifications. It beats the best known turbo codes, at least for bit error rates above 10^{-5}, making it the best error correcting code of rate 1/4 for the Gaussian channel currently known. Not only is the error-correction performance better than that of the turbo code, the block length is less than that of the turbo code. Another key difference between LDPC codes and turbo codes is that, empirically, all errors made by the LDPC decoding algorithm are detected errors. That is, the decoder reports the fact that a block has been incorrectly decoded.

Recent results by Richardson, Shokrollahi and Urbanke [178] have shown that extremely long block length (10^6 bits) irregular LDPC codes can perform within 0.1dB of the Shannon limit (see Figure 9.11). Empirical results were presented for rate 1/2 codes.

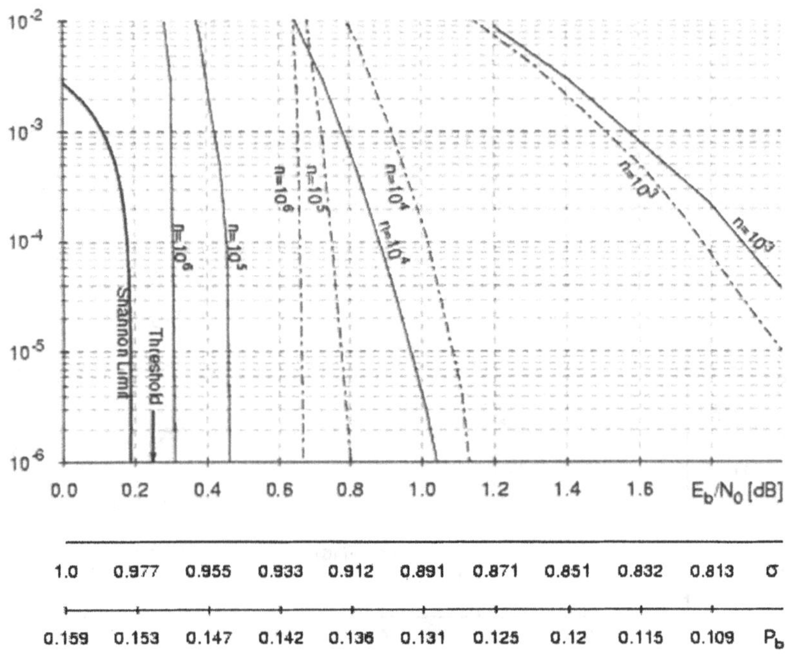

Figure 9.11. Comparison between Turbo Codes (dashed curves) and LDPC Codes (solid curves) of Lengths $n = 10^3, 10^4, 10^5$ *and* 10^6. All codes are of rate one-half. Observe that longer LDPC codes outperform turbo codes and that the gap becomes the more significant with larger n. For short lengths it appears that the structure in turbo codes gives them an edge over LDPC codes despite having a lower threshold.

9.6.2 Analysis of LDPC Codes Performance

The analysis of a low-density code of long block length is difficult because of the immense number of codewords involved. It is simpler to analyze a whole ensemble of such codes because the statistics of an ensemble permit one to average over quantities that are not tractable in individual codes. From the ensemble behavior, one can make statistical statements about the properties of the member codes. Furthermore, one can with high probability find a code with these properties by random selection from the ensemble.

For a wide variety of channels, the Noisy Channel Coding Theorem of Information Theory proves that if properly coded information is transmitted at a rate below the channel capacity, then the probability of decoding error can be made arbitrarily small with the increase of the code length. The theorem does not, however, relate the code length to the computation time or the equipment cost necessary to achieve this low error probability.

The minimum distance of a code is the number of positions in which the two nearest codewords differ. Over the ensemble, the minimum distance of a member code is a random variable, and it can be show that the distribution function of this random variable can be upper bounded by a function such as the one sketched in Figure 9.12 [162]. As the block length increases, for fixed $p \geq 3$ and $q > p$, this function approaches a unit step at a fixed fraction δ_{pq} of the block length. Thus, for large n, practically all the codes in the ensemble have a minimum distance of at least $n\delta_{pq}$. In Table 9.1 [162], this ratio of typical minimum distance to block length is compared to that for a parity-check code chosen at random, *i.e.*, with a matrix filled in with equiprobable independent binary digits. It should be noted that for all the specific nonrandom procedures known for constructing codes, the ratio of the minimum distance to block length appears to approach 0 with increasing block length.

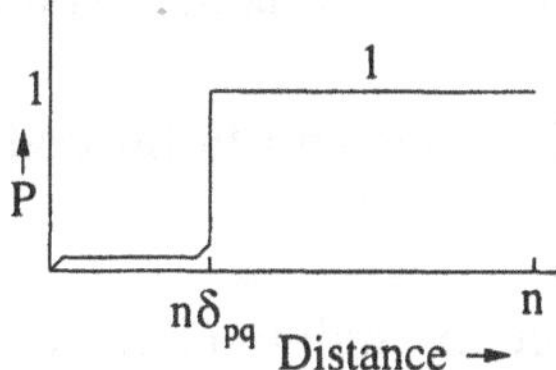

Figure 9.12. Sketch of Bound to Minimum Distance Distribution Function

p	q	*Rate*	δ_{pq}	δ
5	6	0.167	0.255	0.263
4	5	0.2	0.210	0.241
3	4	0.25	0.122	0.214
4	6	0.333	0.129	0.173
3	5	0.4	0.044	0.145
3	6	0.5	0.023	0.11

Table 9.1. Comparison of δ_{pq}, the Ratio of Typical Minimum Distance to Block Length for an (n, p, q) Code, to δ, the Same Ratio for an Ordinary Parity-check Code of the Same Rate.

Although this result for the BSC shows how closely low-density codes approach the optimum, the codes are not designed primarily for use on this channel. The BSC is an approximation to physical channels only when there is a receiver that makes decisions on the incoming signal on a bit-by-bit basis. Since the decoding procedures described earlier actually use the channel a posteriori probabilities, and since a bit-by-bit decision throws away available information, we are actually interested in the probability of decoding error of a binary-input, continuous-output channel. If the noise affects the input symbols symmetrically, then this probability can again be bounded by an exponentially decreasing function of the block length, but the exponent is a rather complicated function of the channel and code. It is expected that the same type of result holds for a wide class of channels with memory, but no analytical results have yet been derived.

9.7. Summary

In this Chapter, the original LDPC code and its variants are introduced, along with the decoding procedure. The first description of an iterative decoding algorithm was by Gallager in 1962, for his low-density parity-check codes that have a simple description and a largely random structure. MacKay [14] proved that sequences of low-density parity-check codes exist that, when decoded with an optimal decoder, approach arbitrarily close to the Shannon limit. The iterative decoding algorithm makes decoding practical and is capable of near Shannon limit performance.

Low-density parity-check codes and turbo codes have several features in common:

- Both have a strong pseudo-random element in their construction
- Both can be decoded using an iterative belief propagation algorithm
- Both have shown to achieve near Shannon limit error-correction performance

Low-density parity-check codes are also shown to be useful for communicating over channels which make insertions and deletions as well as additive (substitution) errors. Error-correction for such channels has not been widely studied, but is of importance whenever synchronization of sender and receiver is imperfect. Davey [174] introduced concatenated codes using novel non-linear inner codes that he called 'watermark' codes, and LDPC codes over non-binary fields as outer codes. The inner code allows resynchronization using a probabilistic decoder, providing soft outputs for the outer LDPC decoder. Error-correction performance using watermark codes is several orders of magnitude better than any comparable results in the literature.

Appendix: The Contents of CD-ROM

In the attached CD-ROM, we have included programs for simulating different Turbo Coding systems. In each case, we have included the end-to-end system including the source, encoder, channel and the decoder. The programs, in the CD-ROM are:

1 TCC

- TCC_Binary_UNIX.c : This program simulates the binary turbo convolutional code given in [7] and explained in Chapter 2 for the UNIX environment.
- TCC_Binary_PC.c : This program simulates the binary turbo convolutional code given in [7] and explained in Chapter 2 for the Windows environment.
- TCC_DVB-RCS_UNIX.c : This program simulates DVB-RCS turbo code given in Chapter 3 for the UNIX environment.
- TCC_DVB-RCS_PC.c : This program simulates DVB-RCS turbo code given in Chapter 3 for the Windows environment.
- TCC-3GPP_UNIX.cpp : This program simulates the 3GPP turbo code given in Chapter 2 for the UNIX environment.
- TCC-3GPP_PC.cpp : This program simulates the 3GPP turbo code given in Chapter 2 for the Windows environment.

2 BTC

- BTC_General_UNIX.c : This program simulates $(n,\ k)^2$ Reed-Muller turbo code for different values of n and k for the UNIX environment.
- BTC_ATM_UNIX.c : This program simulates the shortened $(32,\ 26)^2$ RM codes with different sizes ≤ 676 bits. This program can be, in particular, useful for designing codes for applications with ATM size cells for the UNIX environment.

3 Header Files

- ECHELON.h
- Encoder_RM64.h
- RM_64.h

Each program contains enough information concerning the choice of parameters and options. The authors appreciate receiving feedback from readers concerning the contents of the book and the programs. You may send your comments to y_gao@ece.concordia.ca

References

[1] C.E. Shannon, A Mathematical Theory of Communication, *Bell System Technical Journal*, 27, Part I: 379–423, July 1948. Part II: 623–656, Oct. 1948.

[2] J.T. Coffey, and R.M. Goodman, Any Code of Which We Cannot Think is Good, *IEEE Trans. Inform. Theory*, IT-36(6):1453–1461, November 1990,

[3] G. Battail, We Can Think of Good Codes and Even Decode Them, in *Eurocode'92*, Oct. 26-30, 1992, Udine Italy, printed in the *CISM Courses and lectures*, No. 339, pp. 353-368, Springer 1993.

[4] J. Lodge, P. Hoeher, and J. Hagenauer, The Decoding of Multidimensional Codes Using MAP 'Filters', in *Proc. of 16th Biennial Symp. on Commun.*, Queen's University, Kingston, Ontario, Canada, pp. 343–346, May 1992,

[5] J. Lodge, R. Young, P. Hoeher, and J. Hagenauer, Separable MAP 'Filters' for the Decoding of Product and Concatenated Codes, *Proc. of the 1993 Int. Conf. on Commun., ICC1993*, Geneva, Switzerland, pp. 1740–1745, May 1993.

[6] C. Berrou, A. Glavieux, and P. Thitimajshima, Near Shannon Limit Error-correcting Coding and Decoding: Turbo Codes, *Proc. of the 1993 Int. Conf. on Commun., ICC1993*, pp. 1064–1070, Geneva, Switzerland, May 1993.

[7] J. Hagenauer, E. Offer, and L. Papke, Iterative Decoding of Binary Block and Convolutional Codes, *IEEE Trans. Inform. Theory,* Vol. 42, No. 2, pp. 429-445, March 1996.

[8] R.J. McEliece, On the BCJR Trellis for Linear Block Codes, *IEEE Trans. Inform. Theory*, Vol. 42, No. 4, pp. 1072-1092, July 1996.

[9] C. Berrou, and A. Glavieux, Near Optimum Limit Error Correcting Coding and Decoding: Turbo Codes, *IEEE Trans. Commun.*, Vol. 44, No. 10, pp. 1261-1271, October 1996.

[10] R.M. Pyndiah, Near Optimum Decoding of Product Codes: Block Turbo Codes, *IEEE Trans. Commun.*, Vol. 46, No.6, pp. 1003-1010, August 1998.

[11] S. Benedetto, D. Divsalar, G. Montorsi, and F. Pollara, Serial Concatenation of Interleaved Codes: Performance Analysis, Design, and Iterative

Decoding, *IEEE Trans. Inform. Theory*, Vol. 44, No. 3, pp. 909-926, May 1998.

[12] P. Jung, Comparison of Turbo-code Decoders Applied to Short Frame Transmission Systems, *IEEE Journal on Selected Areas in Communications,* Vol. 14, pp. 530-537, April 1996.

[13] M.C. Davey and D.J.C. MacKay, Low Density Parity Check Codes over GF(q), *IEEE Commu. Lett.,* Vol. 2, No. 6, pp. 165-167, June 1998.

[14] D.J.C. MacKay, Good Error-Correcting Codes Based on Very Sparse Matrices, *IEEE Trans. Inform. Theory,* Vol. 45, No. 2, pp. 399-431, March 1999.

[15] R.J. McEliece, Are Turbo-like Codes Effective on Nonstandard Channels?, *2001 ISIT Plenary Lecture, printed in the IEEE Information Theory Newsletter*, Vol. 51, No. 4, pp. 1-8, Dec. 2001.

[16] E. Boutillon, J. Castura and F.R. Kschischang, Decoder-First Code Design, in *Proc. of the 2nd. Int. Symp. on Turbo Codes and Related Topics*, pp. 459-462, Brest, France, Sept. 2000.

[17] C. Berrou, Turbo Codes: Some Simple Ideas for Efficient Communications, *7th Int. Workshop on Digital Signal Processing Techniques for Space Communications (DSP 2001)*, Sesimbra, Portugal, October 1-3, 2001.

[18] G.D. Forney, The Viterbi Algorithm, *Proceedings of IEEE*, pp. 268-278, March 1973.

[19] K. Chugg, A. Anastasopoulos and Xiapeng Chen, *Iterative Detection: Adaptivity, Complexity Reduction, and Applications*, Kluwer Academic Publishers, 2001.

[20] J.G. Proakis and M. Salehi, *Communication Systems Engineering*, Prentice Hall, 2000.

[21] S. Haykin, *Communication Systems*, 4th. edition, John Wiley & Sons, 2001.

[22] A.J. Viterbi and J.K. Omura, *Principles of Digital Communications and Coding*, McGraw-Hill Book Company, 1979.

[23] G. Ungerboeck, Channel Coding with Multilevel/Phase Signals, *IEEE Trans. Inform. Theory*, Vol. IT-28, pp. 55-68, January 1982.

[24] R.W. Hamming, Error Detecting and Error Correcting Codes, *BellSyst. Tech. J.,* Vol. 29, pp. 147-160, April 1950.

[25] A. Hocquenghem, Codes Correcteurs d'Erreurs, *Chiffres,* Vol. 2, pp. 147-156, 1959.

[26] R.C. Bose and D.K. Ray-Chaudhuri, On a Class of Error Correcting Group Code, *Inf. Control,* Vol. 3, pp. 68-79, March 1960.

[27] I.S. Reed and G. Solomon, Polynomial Codes over Certain Finite Fields, *J. Soc. Ind. Appl. Math.*, Vol. 8, pp. 300-304, June 1960.

[28] *Digital Video Broadcasting (DVB) interaction channel for satellite distribution system*, ETSI reference EN 301 799, v1.2.2, Dec. 2000.

[29] A.J. Viterbi, Error Bounds for Convolutional Codes and an Asymptotically Optimum Decoding Algorithm, *IEEE Trans. Inform. Theory*, Vol. IT-13, pp. 260-269, April 1967.

[30] R.G. Gallager, *Information Theory and Reliable Communication*, John Wiley, 1968.

[31] Richard E. Blahut, *Principles and Practice of Information Theory*, Addison-Wesley Publishing Company, 1987.

[32] *Forward Error Correction Data Book*, QUALCOMM Inc., ASIC Products, 80-24128-1A, 8/98.

[33] J.P. Odenwalder, *Optimal Decoding of Convolutional Codes*, Ph.D. Thesis, University of California, Los Angeles, 1970.

[34] Shu Lin and Daniel J. Costello, Jr., *Error Control Coding: Fundamentals and Applications*, Prentice Hall, Inc., Englewood Cliffs, New Jersey, 1983.

[35] H. Nickl, J. Hagenauer, and F. Burkert, Approaching Shannon's Capacity Limit by 0.27 dB using Hamming Codes in a Turbo-decoding Scheme, in *Proc. IEEE Intl. Symposium on Information Theory*, June 1997.

[36] C. Douillard *et al.* The Turbo Code Standard for DVB-RCS, in *Proc. of the 2nd Int Symp. on Turbo codes*, Brest, France, pages 551-554, Sept., 2000.

[37] European Telecommunications Standards Institute (ETSI) TS 125 212, Universal Mobile Telecommunications System (UMTS); Multiplexing and Channel Coding (FDD), *3GPP TS 25.212 version 5.0.0 Release 5*, Mar. 2002.

[38] M.Eroz and A.R.Hammons, On the Design of Prunable Interleavers for Turbo Codes, in *Proc. Vehicular Technology Conference*. Houston, USA, pp. 1669-1673, May 1999.

[39] 3rd Generation Partnership Project 2 (3GPP2) CDMA2000 High Rate Packet Data Air Interface Specification, *3GPP2 C:S0024 version 2*, Aug. 2001.

[40] Yufei Wu Design and Implementation of Parallel and Serial Concatenated Convolutional Codes, Ph.D. dissertation, Virginia Polytechnic Institute and State University, June 1999.

[41] P. Elias, Error-Free Coding, *IRE Trans. Inform. Theory*, pp. 29-37, Sept. 1954.

[42] S.B. Wicker, *Error Control Systems for Digital Communication and Storage*, Prentice-Hall Englewood Cliffs, 1995.

[43] J.G. Proakis, *Digital Communications*, New York: McGraw-Hill,Inc., third ed., 1995.

[44] J.G.Proakis, *Digital Communications*, New York: McGraw-Hill,Inc., Fourh Edition, 2001.

[45] I.S. Reed, A Class of Multiple-Error-Correcting Codes and a Decoding Scheme, *IEEE Trans. Inform. Theory*, pages 38-49, September, 1954.

[46] G.D. Forney, *Concatenated Codes*, Cambridge, MA: MIT Press, 1966.

[47] P. Elias, Coding for Noisy Channels, *IRE Conv. Rec., Part 4*, pages 37–47, 1955.

[48] J.M. Wozencraft and B. Reiffen, *Sequential Decoding*, Cambridge, MA: MIT Press, 1961.

[49] J.L. Massey, *Threshold Decoding*, Cambridge, MA: MIT Press, 1963.

[50] J.K. Omura, On the Viterbi Decoding Algorithm, *IEEE Trans. Inform. Theory*, IT-15, pages 177–179, January 1969.

[51] G.D. Forney, Convolutional Codes II: Maximum Likelihood Decoding, *Inf. Control*, Vol. 25, pages 222-266, July, 1974.

[52] Consultative Committee for Space data Systems, Recommendations for Space Data Standard: Telemetry Channel Coding, Blue Book Issue 2, CCSCS 101.0-B2, Jan., 1987.

[53] J. Hagenauer, E. Offer and L. Papke, Matching Viterbi Decoders and Reed-Solmon Decoders in Concatenated Systems, *Reed-Solomon Codes and their Applications*, (S.B.Wicker and V.K.Bhargava,eds.), Piscataway, NJ: IEEE press, pages 242-271, 1994.

[54] J. Hagenauer and P. Hoeher, A Viterbi Algorithm with Soft-Decision Outputs and its Applications, in *Proc. IEEE GLOBECOM'89*, pages 1680-1686, 1989.

[55] Jun Tan and Gordon L. Stuber, A MAP Equivalent SOVA for Non-binary Turbo Codes, in *Proc. IEEE ICC'89*, New Orleans, LA, pages 602-606, June 2000.

[56] P. Robertson, E. Villebrun and P. Hoeher, A Comparison of Optimal and Sub-optimal MAP Decoding Algorithms Operating in the Log Domain, in *Proc. IEEE ICC'95*, New Orleans, LA,Seattle, WA, pages 1009-1013, June 2000.

[57] L. Bahl, J. Cocke, F. Jelinek and J. Raviv, Optimal Decoding of Linear Codes for Minimizing Symbol Error Rate, *IEEE Trans. Inform. Theory*, Vol.IT-20, pages 284-287, March, 1974.

[58] D.Divsalar and F.Pollara, Serial and Hybrid Concatenation Codes with Applications, in *Proc. Int Symp. on Turbo codes and Related Topics*, Brest, France, pages 80-87, Sept., 1997.

[59] S. Benedetto, G. Montorsi, D. Divsalar and F. Pollara, Serial Concatenation of Interleaved Codes: Performance Analysis, Design, and Iterative Decoding, *JPL TDA Progress Report*, pages 42-126, Aug. 1996.

[60] B. Vucetic and Jinhong Yuan, *Turbo Codes Principles and Applications*, Kluwer Academic Publishers, Boston/Dordrecht/London, 2000.

[61] F. Berens, A. Worm, H. Michel and N. Wehn, Implementation Aspects of Turbo-Decoders of Future Radio Applications, in *Proc. VTC'99*, Amsterdam, Netherlands, pages 2601-2605, Sept. 1999.

[62] Roberto Garello, Paola Pierleoni and Sergio Benedetto, Computing the Free Distance of Turbo Codes and Serially Concatenated Codes with Interleavers: Algorithms and Applications, *IEEE Journal on Selected Areas in Communications*, Vol. 19, No. 5, pages 800-812, May, 2001.

[63] Johan Hokfelt, Ove Edfors and Torleiv Maseng, On the Theory and Performance of Trellis Termination Methods for Turbo Codes, *IEEE Journal on Selected Areas in Communications*, Vol. 19, No. 5, pages 838-847, May, 2001.

[64] J. Andersen and V. Zyablov, Interleaver Design for Turbo Coding, in *Proc. Int. Symp. on Turbo Codes and Related Topics*, Brest, France, Sept. 1997.

[65] S. Crozier, J. Lodge, P. Guinand and A. Hunt, Performance of Turbo Codes with Relative Prime and Golden Interleaving Strategies, in *Proc. Sixth Int. Mobile Satellite Conf.*, Ottawa, Canada, pages 268-275, June, 1999.

[66] F. Daneshgaran and M. Mondin, Design of Interleavers for Turbo BCdes Based on a Cost Function, in *Proc. Int. Symp. on Turbo Codes and Related Topics*, Brest, France, pages 255-258, Sept., 1997.

[67] S. Dolinar and D. Divsalar, Weight Distributions for Turbo Codes using Random and Nonrandom Permutations, TDA progress report, Jet propulsion Lab., Pasadena, CA, pages 42-122, Aug. 1995.

[68] A.S. Barbulescu and S.S. Pietrobon, Interleaver Design for Three Dimensional Turbo-Codes, in *Proc. IEEE Int. Symp. on Inform. Theory*. Whistler, BC, Canada, Sept. 1995.

[69] M. Hattori, J. Murayama and R.J. McEliece, Psedo-Random and Self-Terminating Interleavers for Turbo Codes, *Inform. Theory Workshop*, pages 9-10, Feb. 1998.

[70] J. Hokfelt, O. Edfors and T. Maseng, Interleaver Structures for Turbo Codes with Reduced Storage Memory Requirement, in *Proc. Vehicular Technology Conference*, Amsterdam, Netherlands, pages 1585-1589, Sept. 1999.

[71] J. Hokfelt and T. Maseng, Methodical Interleaver Design for Turbo Codes, in *Proc. Int. Symp. on Turbo Codes and Related Topics*. Brest, France, pages 212-215, Sept. 1997.

[72] H.R. Sadjapour, M. Salehi, N.J.A. Sloane and G. Nebe, Interleaver Design for Short Block Length Turbo Codes, *Proc. IEEE Int. Conf. on Commun.*, New Orleans, USA, June, 2000.

[73] M. Oberg, Turbo Coding and Decoding for Signal Transmission and Recording Systems, PhD thesis, University of California,San Diego, CA, USA, 2000.

[74] A. Shibutani, H. Suda and F. Adachi, Multi-Stage Interleaver for Turbo Codes in DS-CDMA Mobile Radio, in *Proc. Asia-Pasific Conference on Communications*, Nov., 1998.

[75] O.Y. Takeshita and D.J. Costello, New Classes of Algebraic Interleavers for Turbo-Codes, in *Proc. IEEE int. Symp. on Inform. Theory*, Cambridge, MA, USA, page 419, Aug., 1998.

[76] H.R. Sadjapour, M. Salehi, N.J.A. Sloane and G. Nebe, Inerleaver Design for Turbo Codes, *IEEE Journal on Selected Areas in Communication*, Vol.19, No.5, May 2001.

[77] W. Blacker, E. Hall and S. Wilson, Turbo Code Termination and Interleaver Conditions, *IEEE Electron. Lett.*, Vol. 31, No. 24, pages 2082-2084, Nov., 1995.

[78] A.S. Barbulescu and S.S. Pietrobon, Terminating the Trellis of Turbo Codes in the Same State, *IEEE Electron. Lett.*, Vol. 31, No. 1, pages 22-23, Jan., 1995.

[79] M.C. Reed and S.S. Pietrobon, Turbo Code Termination Schemes and a Novel Alternative for Short Frames, in *Proc. 7th IEEE Int. Symp. Personal, Indoor, Mobile Communications*, Taipai, Taiwan, Vol. 2, pages 354-358, Oct., 1996.

[80] P. Guinand and J. Lodge, Trellis Termination for Turbo Encoders, in *Proc. 17th Biennial Symp. on Commun.*, Kingston, Canada, pages 389-392, May, 1994.

[81] J. Hokfelt, O. Edfors and T. Maseng, A Survey on Trellis Termination Alternatives for Turbo Codes, in *Proc. IEEE Vehicular Technology Conference (VTC'99)*, Houston, Texas, USA, pages 2225-2229, May, 1999.

[82] J. Hokfelt, C.F. Leanderson, O. Edfors and T. Maseng, Distance Spectrum of Turbo Codes using Different Trellis Termination Methods, in *Proc. Int. Sym. on Turbo Codes and Related Topics*, Brest, France, pages 463-466, Sept., 1997.

[83] J.B. Anderson and S.M. Hladik, Tailbiting MAP Decoders, *IEEE Journal on Selected Areas in Communications*, Vol. 16, No. 2, pages 297-302, Feb., 1998.

[84] S. Crozier, P.Guinand, J. Lodge and A. Hunt, Construction and Performance of New Tail-Biting Turbo Codes, *6-th International Workshop on Digital Signal Processing Techniques for Space Applications (DSP'98)*, Noordwijk, Netherlands, Sept., 1998.

[85] N.A. Van Stralen, J.A.F. Ross and J.B. Anderson, Tailbiting and Decoding Recursive Systematic Codes, *IEEE Electron. Lett.*, Vol. 35, No. 17, pages 1461-1462, Aug., 1999.

[86] Y.P. Wand, R. Ramesh, A. Hassan and H. Koorapaty, On MAP Decoding for Tail-Biting Convolutional Codes, in *Proc. IEEE Int. Symp. on Inform. Theory*, page 225, June, 1997.

[87] C. Weib, C. Bettstetter S. Riedel and D.J. Costello, Turbo Decoding with Tail-Biting Trellises, in *Proc. URSI Int. Symp. on Signals, Systems and Electronics*, pages 343-348, Sept., 1998.

[88] C. Berrou, C. Douillard and M. Jezequel, Designing Turbo Codes for Low Rates, *Digest of IEE Colloq. on "Turbo Codes in Digital Broadcasting-could it Double Capacity?"*, Vol. 165, Nov., 1999.

[89] C. Berrou and M. Jezequel, Non Binary Convolutional Codes for Turbo Coding, *IEEE Electronic Letters*, Vol. 35, No. 1, pages 39-40, Jan., 1999.

[90] C. Berrou, C. Douillard and M. Jezequel, Multiple Parallel Concatenation of Circular Recursive Convolutional (CRSC) Codes, *Annals of Telecommunications*, Vol. 54, No. 3-4, pages 166-172, March-April, 1999.

[91] C. Berrou and A. Glavieux, Turbo Codes, General Principles and Applications, in *Proc. of the 6th Int. Tirrenia Workshop on Digital Communications*, Pisa, Italy, pages 215-226, Sept., 1993.

[92] N. Brengarth, R. Novello, N. Pham, V. Piloni and J. Tousch, DVB-RCS Turbo Code on a Commercial OPB Satellite Payload: Skyplex, in *Proc. of the 2nd Int. Symp. on Turbo codes*, Brest, France, pages 535-538, Sept., 2000.

[93] S. Dolinar, D. Divsalar and F. Pollara, Code Performance as a Function of Block Size, TMO progress report, JPL, NASA, pages 42-133

[94] Y. Wu and B.D. Woerner, The Influence of Quantization and Fixed Point Arithmetic upon the BER Performance of Turbo Codes, in *Proc. IEEE International Conference on Vehicular Technology (VTC'99)*, Vol. 2, pages 1683-1687, May, 1999.

[95] Y. Wu and B.D. Woerner, Internal Data Width SISO Decoding Module with Modular Renormalization, in *Proc. IEEE Veh. Tech. Conf.*, Tokyo, Japan, May, 2000.

[96] D.E. Cress and W.J. Ebel, Turbo Code Implementation Issues for Low Latency, Low Power Applications, in *Proc. Symp. on wireless Personal Communications*, MPRG, Virginia Tech, VA, USA, June, 1998.

[97] Z. Wang *et al.*, VLSI Implementation Issues of Turbo Decoder Design for Wireless Applications, in *Proc. 1999 IEEE Workshop on Signal Processing System (SiPS), Design and Implementation*, Taipei, Taiwan, Oct., 1999.

[98] H. Michel, A. Worm and N. Wehn, Influence of Quantization on the Bit-Error Performance of Turbo-Decoders, in *Proc. IEEE Veh. Tech. Conf. (VTC'00)*, Tokyo, Japan, May, 2000.

[99] H. Michel and Norbert Wehn, Turbo-Decoder Quantization for UMTS, *IEEE Commun. Lett.*, Vol. 5, NO. 2, pp.55-57, Feb. 2001.

[100] Chip Fleming, Simulation Source Code Examples, http://pw1.netcom.com/ chip.f/viterbi.html, 2001.

[101] G. Ungerboeck, Channel Coding with Multilevel/Phase Signals, *IEEE Trans. Inform. Theory*, Vol. IT-28, pages 56-67, Jan., 1982.

[102] G. Ungerboeck, Trellis-Coded Modulation with Redundant Signal Sets, Part I: Introduction, *IEEE Commun. Mag.*, Vol. 25, No. 2, 1987.

[103] G. Ungerboeck, Trellis-Coded Modulation with Redundant Signal Sets, Part II: State of the Art, *IEEE Commun. Mag.*, Vol. 25, No. 2, pages 12-21, 1987.

[104] Patrick Robertson and Thomas Worz, Coded Modulation Scheme Employing Turbo Codes, *IEEE Electron. Lett.*, Vol. 31, pages 1546-1547, Aug., 1995.

[105] Prof. Nandana Rajatheva's homepage, Turbo Codes with High Spectral Efficiency, *http://www.ucop.edu/research/micro/99_00/99_123.pdf.*

[106] G.Y. Liang *et al.*, High Performance 3GPP Turbo Decoder Implemented ON Texas Instrument ? TMS320C6201 ? DSP, *http://www.ntu.edu.sg/ntrc/personalcomm/high_performance_3gpp_turbo_deco.htm.*

[107] H. Imai and S. Hirakawa, A New Multilevel Coding Method using Error Correcting Codes, *IEEE Trans. Inform. Theory*, Vol. 23, No. 3, pages 371-377, May 1977.

[108] Patrick Robertson and Thomas Worz, Bandwidth-Efficient Turbo Trellis-Coded Modulation Using Punctured Component Codes, *IEEE Journal on Selected Areas in Communications*, Vol. 16, No. 2, pages 206-218, Feb., 1998.

[109] S. LeGoff, A. Glavieux and C. Berrou, Turbo Codes and High Efficiency Modulation, *Proc. of IEEE ICC'94*, New Orleans, LA, pages 645-649, May 1994.

[110] Yingzi Gao, Design and Implementation of Non-binary Convolutional Turbo Code, M.A.Sc. thesis, Dept. of Elect. & Comp. Eng., Concordia University, Dec. 2001.

[111] Yingzi Gao and M.R.Soleymani, Spectrally Efficient Non-binary Turbo Codes: Beyond DVB-RCS standard, in *Proceedings of 21th Biennial Symposium on Communications*, pp. 5-9, Queen's University, Kingston, Ontario, Canada, May 2002,

[112] K. Gracie, S. Crozier, A. Hunt and J. Lodge, Performance of a Low-Complexity Turbo Decoder and its Implementation on a Low-Cost, 16-Bit Fixed-Point DSP, in *Proc. of the 10th Int. Conf. on Wireless Commun. (Wireless'98)* Calgary, AB, Canada, pp.229-238, Jul. 1998.

[113] K. Gracie, S. Crozier and A. Hunt, Performance of a Low-Complexity Turbo Decoder with a Simple Early Stopping Criterion Implemented on a SHARC Processor, *Sixth Int. Mobile Satellite Conf. (IMSC'99)* Ottawa, Canada, pp.281-286, June 1999.

[114] Ken Gracie, Personal Correspondence.

[115] S. Crozier, K. Gracie and A. Hunt, Efficient Turbo Decoding Techniques, in *Proc. Int. Conf. Wireless Commun. (Wireless'99)*, Calgary, Canada, July 1999.

[116] A. Hunt, S. Crozier, M. Richards and K. Gracie, Performance Degradation as a Function of Overlap Depth when using Sub-block Processing in the Decoding of Turbo Codes, in *Proc. Sixth Int. Mobile Satellite Conf. (IMSC'99)*, Ottawa, Canada, pp. 276-280, June 1999.

[117] J.P.Woodard, Implementation of High Rate Turbo Decoders for Third Generation Mobile Communications, *IEE Colloquium on "Turbo Codes in Digital Broadcasting-Could It Double Capacity?"* pp.12/1-12/6, Nov. 1999.

[118] C. Berrou, *et al.*, The Advantages of Non-Binary Turbo Codes, in *Proc. IEEE ITW2001*, Caims, Australia, pp. 61-63, Sept. 2001.

[119] P. Robertson, E. Villebrun, and P. Hoeher, A Comparison of Optimal and Sub-Optimal MAP Decoding Algorithms Operating in the Log Domain, in *Proc. IEEE Int. Conf. on Commun.*, Seattle, WA, pp. 1009-1013, June 1995.

[120] Y. Liu, H. Tang, M. Fossorier and S. Lin, Iterative Decoding of Concatenated Reed-Solomon Codes, 37th *Annual Allerton. Conf.*, Sept. 1999.

[121] Y. Liu, S. Lin, and M. Fossorier, MAP Algorithm for Decoding Linear Block Codes Based on Sectionalized Trellis Diagrams, *IEEE Trans. Inform. Theory*, Vol. 48, No. 4, April 2000.

[122] D. Chase, A Class of Algorithm for Decoding Block Codes with Channel Measurement Information, *IEEE Trans. Inform. Theory*, Vol. IT-18, pp. 170-182. Jan. 1972

[123] Advanced Hardware Architectures (AHA), PS4501: Astro 36 Mbits/s Turbo Product Code Encoder/Decoder.

[124] F. Buda, J. Feng and P. Sehier, Soft Decoding of BCH Codes Applied to Multilevel Modulation Codes for Rayleigh Fading Channels, in *Proc. MILCOM97*, New York, USA. Vol. 1, pp. 32–36, 1997.

[125] J. Feng. F. Buda, A Special Family of Product Codes "Turboly" Decodable with Application to ATM Cell Transmission, in *IEEE Int. Symp. on Inform. Theory*, New York, USA, pp. 289, 1998.

[126] J. Fang, F. Buda and E. Lemois, Turbo Product Code: A Well Suitable Solution to Wireless Packet Transmission for Very Low Error Rates, in *Proc. Int. Symp. on Turbo Codes and Related Topics*, Brest, France, pp. 101-111, Sept. 2000.

[127] A. Berthet, J. Fang and P. Tortelier, Generalized Turbo Product Codes and their Properties in Iterative SISO Decoding, in *Proc. Int. Symp. on Turbo Codes and Related Topics*, Brest, France, pp. 499-502, Sept. 2000.

[128] R. Pyndiah, A. Glavieux, A. Picart, and S. Jacq, Near Optimum Decoding of Product Codes, *Proc. IEEE GLOBECOM*, San Francisco, USA, pp. 339–343, Nov. 1994.

[129] R. Pyndiah, Pierre Combelles and P. Adde, A Very Low Complexity Block Turbo Decoder for Product Codes, *Proc. IEEE GLOBECOM*, London, pp. 101-105, Nov. 1996.

[130] O. Aitsab, R. Pyndiah, Performance of Reed Solomon Block Turbo Codes, in *Proc IEEE GLOBECOM*, London, UK, pp. 121-125, Nov. 1996.

[131] P. Adde and R. Pyndiah, Recent Simplifications and Improvements in Block Turbo Codes, in *Proc. Int. Symp. on Turbo Codes and Related Topics*, Brest, France, pp. 133-136, Sept. 2000.

[132] S. A. Hirst, B. Honary and G. Markarian, Fast Chase Algorithm with Application in Turbo Decoding, in *Proc. Int. Symp. on Turbo Codes and Related Topics*, Brest, France pp. 259-262, Sept. 2000.

[133] S. Kerouedan and P. Adde, Implementation of a Block Turbo Decoder on a Single Chip, in *Proc. Int. Symp. on Turbo Codes and Related Topics*, Brest, France, pp.243-246, Sept. 2000.

[134] A. Goalic and N. Chapalain, Real Time Turbo Decoding of BCH Product Code on the DSP Texas TMS320C6201, in *Proc. Int. Symp. on Turbo Codes and Related Topics*, Brest, France, pp. 331-334, Sept. 2000.

[135] M. Vanderaar, R. T. Gedney and E. Hewitt, Comparative Performance of Turbo Product Codes and Reed-Solomon/Convolutional Concatenated Codes for ATM Cell Transmission, *Fifth Ka Band Utilization Conf.*, Toarmina, Italy, October 1999.

[136] S. B. Wicker, *Error Control Systems for Digital Communication and Storage*, Prentice-Hall, Englewood Cliffs, NJ, USA, 1995.

[137] J. K. Wolf, Efficient Maximum-Likelihood Decoding of Linear Block Codes, *IEEE Trans. Inform. Theory*, Vol. IT-24, pp. 76–80, Jan. 1978.

[138] G.D. Forney, Coset codes II: Binary Lattices and Related Codes, *IEEE Trans. Inform. Theory*, Vol. 34, No. 5, pp. 1152-1187, Sept. 1988.

[139] D. J. Muder, Minimal Trellises for Block Codes, *IEEE Trans. Inform. Theory*, Vol. 34, No. 5, pp. 1049-1053, Sept 1988.

[140] Y. Berger and Y Be'ery, The Twisted Squaring Construction Trellis Complexity and Generalized Weights of BCH and QR codes, *IEEE Trans. Inform. Theory*, Vol. 42, No. 6, pp.1817-1827, Nov. 1996.

[141] F. R. Kschischang and V. Sorokine, On the Trellis Structure of Block Codes, *IEEE Trans. Inform. Theory*, Vol. 41, No. 6, pp. 1924-1937, Nov. 1995.

[142] R. J. McEliece, On the BCJR Trellis for Linear Block Codes, *IEEE Trans. Inform. Theory*, Vol. 42, No. 4, pp. 1072-1092, July 1996.

[143] G. Horn and F. R. Kschischang, On the Intractability of Permuting a Block Code to Minimize Trellis Complexity, *IEEE Trans. Inform. Theory*, Vol. 42, No. 6, pp. 2042-2048, Nov. 1996.

[144] T. Kasami, T. Takata, T. Fujiwara, and S. Lin, On the State Complexity of Trellis Diagrams for Reed-Muller Codes and their Supercodes, *Proc. 14th Symp. on Inform. Theory and Its Applications*, Ibusuki, Japan, pp. 101-104, Dec. 1991.

[145] J. L. Massey, Foundation and Methods of Channel Encoding, *Proc. Int. Conf. Inform. Theory and Systems*, vol. 65, NTG-Fachberichte, Berlin, pp. 148-157, 1978.

[146] V. S. Pless and W. C. Huffman, *Editors, Handbook of Coding Theory*, Elsevier Science B.V., Volume II, Amsterdam, Netherlands, pp. 1989-2117, 1998.

[147] A. M. Michelson and A. H. Levesque, *Error-Control Techniques for Digital Communication*, John Wiley & Sons, 1985.

[148] M. Vanderaar, Efficient Channel Coding (ECC) Inc., Personal correspondence.

[149] A. Giulietti, J. Liu, F. Maessen, A. Bourdoux, L. van der Perre, B. Gyselinckx, M. Engels, M. Strum, A Trade-Off Study on Concatenated Channel Coding Techniques for High Data Rate Satellite Communications, in *Proc. Int. Symp. on Turbo Codes and Related Topics*, Brest, France, pp. 125-128, Sept. 2000.

[150] T. A. Summers and S. G. Wilson, SNR Mismatch and Online Estimation in Turbo Decoding, *IEEE Trans. Commun.* , Vol. 46, No.4, April 1998.

[151] W. Oh and K. Cheun, Adaptive Channel SNR Estimation Algorithm for Turbo Decoder, *IEEE Commun. Lett.*, Vol. 4, No. 8, August 2000.

[152] M. S. C. Ho and S. S. Pietrobon, A Variance Mismatch Study for Serial Concatenated Turbo Codes, in *Proc. Int. Symp. on Turbo Codes and Related Topics*, Brest, France, pp. 483-485, Sept. 2000.

[153] IEEE 802.16 Working Group, Local and Metropolitan Area Network Part 16: Standard Air Interface for Fixed Broadband Wireless Access Systems, http://ieee802.org/16/tg3_4/docs/80216ab-01_01r2.zip

[154] Advanced Hardware Architectures (AHA), Applications and Solutions, http://www.aha.com/applications/

[155] Advanced Hardware Architectures (AHA), Next Generation Direct-to-Home Satellite Systems, White paper, http://www.aha.com/technology/showproduct.asp?iId=39

[156] COMTECH EF DATA, Higher Order Modulation and Turbo Coding Options for the Cdm-600 Satellite Modem, Data sheet, http://www.comtechefdata.com

[157] Turbo Concept, TC3000: Turbo Product Code, Data sheet, http://www.turboconcept.com

[158] Paradise Data Com, P300 Turbo Satellite Modem, Data sheet, http://www.paradise.co.uk/products/modems/p300turbo.html

[159] U. Vilaipornsawai, Trellis Based Iterative Decoding of Block Codes for Satellite ATM, M.A.Sc. thesis, Dept. of Elect. & Comp. Eng., Concordia University, Winter 2001.

[160] Bo Yin, Trellis Decoding of 3D Block Turbo Codes, Master thesis, Concordia University, expected Fall 2002.

[161] B.Talibart and C.Berrou, Notice Preliminaire du Circuit Turbo-Condeur/Decodeur TURBO4, *Version 0.0*, June 1995.

[162] R. G. Gallager, Low-Density Parity-Check Codes, *IRE Trans. Inform. Theory*, pp.21-28, Jan 1962.

[163] R. M. Tanner, A Recursive Approach to Low Complexity Codes, *IEEE Trans. Inform. Theory*, Vol. IT-27, pp. 533-547, Sept. 1981.

[164] V. Zyablov and M. Pinsker, Estimation of the Error-Correction Complexity of Gallager Low-Density Codes, *Probl. Pred. Inform.*, Vol. 11, pp. 23-26, Jan. 1975.

[165] G. A. Margulis, Explicit Construction of Graphs without Short Cycles and Low Density Codes, *Combinatorica*, Vol. 2, No. 1, pp. 71-78, Jan. 1982.

[166] N. Wiberg, Codes and Decoding on General Graphs, PhD thesis, Dept. of Electrical Engineering, Linkoping studies in Science and Technology, Dissertation No. 440, Linkoping, Sweden, 1996.

[167] D. J. C. MacKay and R. M. Neal, Good Codes Based on Very Sparse Matrices, *in Cryptography and coding 5th. IMA Conf. C. Boyd, Ed., Lecture Notes in Computer Science*, No. 1025, pp. 100-111, Springer, Berlin, Germany, 1995.

[168] D. J. C. MacKay and R. M. Neal, Near Shannon Limit Performance of Low Density Parity Check Codes, *Electronic Letters*, Vol. 32, pp. 1645-1646, Aug. 1996.

[169] E. R. Berlekamp, R. J. McEliece, and H. C. A. van Tilborg, On the Intractability of Certain Coding Problems, *IEEE Trans. Inform. Theory*, Vol. 24 (3), pp. 384-386, 1974.

[170] Turbo Codes Performance, Available from, http://www331.jpl.nasa.gov/public/TurboPerf.html, August 1996.

[171] M. G. Luby, M. Mitzenmacher, M. A. Shokrollahi, and D. A. Spielman, Improved Low-density Parity-check Codes using Irregular Graphs and Belief Propagation, in *Proc. IEEE Int. Symp. on Inform. Theory (ISIT)*, page 117, 1998.

[172] M. G. Luby, M. Mitzenmacher, M. A. Shokrollahi, and D. A. Spielman, Improved Low-density Parity-check Codes using Irregular Graphs, *IEEE Trans. Inform. Theory*, Vol. 47, No. 2, page 585-598, 2001.

[173] M. C. Davey and D. J. C. MacKay, Low Density Parity Check Codes over GF(q), in *Proc. IEEE Inform. Theory Workshop*, pages 70-71, June 1998.

[174] M. C. Davey, Error-correction using Low-Density Parity-Check Codes, Ph.D. Dissertation, University of Cambridge, Dec. 1999.

[175] D. J. C. MacKay, S. T. Wilson, and M. C. Davey, Low Density Parity Check Codes over GF(q), in *Proc. IEEE Inform. Theory Workshop*, pages 70-71, June 1998.

[176] D. J. C. MacKay, S. T. Wilson, and M. C. Davey, Comparison of Constructions of Irregular Gallager Codes, *IEEE Trans. Commun.*, Vol. 47(10) pp. 1449-1454, Oct. 1999

[177] J. Pearl, Probabilistic Reasoning in Intelligent Systems: Networks of Plausible Inference, Morgan Kaufmann, San Mateo, 1988.

[178] T. Richardson, A. Shokrollahi and R. Urbanke, Design of Provably Good Low-density Parity Check Codes, *IEEE Trans. Inform. Theory*, Vol. 47, No.2, pages 619-637 Feb. 2001.

[179] T. J. Richardson and R. L. Urbanke, The Capacity of Low-density Parity-check Codes under Message-passing Decoding, *IEEE Trans. Inform. Theory*, Vol. 47, No.2, pages 599-618 Feb. 2001.

[180] T. Richardson, A. Shokrollahi and R. Urbanke, Efficient Encoding of Low-Density Parity-Check Codes, *IEEE Trans. Inform. Theory*, Vol. 47, No.2, pages 638-656 Feb. 2001.

Index

Zeitfracht Medien GmbH
Ferdinand-Jühlke-Straße 7
99095 Erfurt, Deutschland
produktsicherheit@kolibri360.de